The New Barbarism and the Modern West

The New Barbarism and the Modern West

Recognizing an Ethic of Difference

Toivo Koivukoski

LEXINGTON BOOKS
Lanham • Boulder • New York • London

Published by Lexington Books
An imprint of The Rowman & Littlefield Publishing Group, Inc.
4501 Forbes Boulevard, Suite 200, Lanham, Maryland 20706
www.rowman.com

16 Carlisle Street, London W1D 3BT, United Kingdom

British Library Cataloguing in Publication Information Available

Library of Congress Cataloging-in-Publication Data

Koivukoski, Toivo.
The new barbarism and the modern West : recognizing an ethic of difference / Toivo Koivukoski.
pages cm
Includes bibliographical references and index.
ISBN 978-0-7391-8999-3 (cloth : alk. paper) -- ISBN 978-0-7391-9000-5 (electronic) 1. Other (Philosophy). 2. Difference (Philosophy) 3. Civilization, Modern--21st century--Philosophy. 4. Technology and civilization. I. Title.
BD460.O74K65 2014
172--dc23
2014021893

∞™ The paper used in this publication meets the minimum requirements of American National Standard for Information Sciences Permanence of Paper for Printed Library Materials, ANSI/NISO Z39.48-1992.

Printed in the United States of America

Contents

Chapter One

Outer Barbarians

AN ETHICAL PROPOSITION

One can understand a civilization at its roots—and perhaps suggest its potential for longevity also—in terms of how it treats its barbarians, that is, how it constitutes itself as civilized in relation to. . . . It is a matter of "its barbarians" and "our" barbarism specifically because the notion of barbarism is quite clearly a culturally dependent term. For with the exception perhaps of self-identifying new primitives recoiling from the lockstep (and at times barbaric) logics of a technological society, people don't generally self-identify as being barbaric themselves. Rather, as a set of social identities, whatever meaning the category has is generated through an imagined projection beyond the margins of a civilizational framework, whereby an identity is formulated in terms of an essential other; that is, the enemies at the gates or the enemy within. What is revealed then in the notion of barbarism as such cannot really pertain to the barbarians themselves but instead demonstrates an attitude of a people toward those with whom they do not identify, along with their own character in the face of collective anxieties.

Yet beyond the vice or virtue of the habitual patterns of behavior embedded in our collective prejudices, there is a certain, almost incidental awareness avowed in the assumptions we make about our others, though in the form of glancing intuition. For if what can come to be known are not those others themselves but rather those blind spots within one's own civilizational worldview, thus to be filled in by proxy, such that if the dominant civilization is not considered hardy enough, or natural enough, then perhaps, it is imagined, these wild, halfway human beings are acting out as the remains of an otherwise hypertrophied civilization. This logic of projected identities informs the constitution of barbaric others, and if it does little to illuminate "the other" as such (that is, others understood in their own terms, in and of themselves and in their own words), it does offer an opportune beginning for inter-cultural relations, as well as for critical self-consciousness, seeing our own limits in our self-made others.

What this means is that one's treatment of others is joined ineluctably to one's own sense of self-respect. That is, the more definitive the self-other distinction is and the sharper the boundaries between the self-accepted self and the cauterized and exiled scraps of the unrealized self, the more incomplete the assumed identity becomes. One must wonder what unconscious reflexes of I against I drive collective identities toward such gestures of self-immolation. On principle, though, I would argue that identity formation by means of opposition and exclusion can never arrive at a robust form of self-knowledge, that is, an identity capable of giving an account of itself. One cannot define a being in terms of what it is not, as if by tracing out all of the other beings around it and thus circumscribing the thing itself at the center of a circle of alterity. One could go on *ad infinitum* listing *not this, . . . not this, . . . not this . . .* without getting any closer to precisely just what *this* is. And crucially so, since otherwise any kind of knowledge of any one thing—or of oneself, for that matter—would require a kind of global experience of all things, with identity vacillating between complete ignorance and total comprehension. Rather, as a middle ground position between universal identity and the abyss of non-identity, what can be traced out are the conscious limits of self-knowledge—that is the points of contact with one's own essential difference. One's own non-being is, in this sense, intrinsic to oneself, such that non-being is essential to one's being. It is in this that we find clues as to our own identities, wherein being is defined not by contrast but in relation to non-being. It is in the gaps or openings within self-consciousness that one can recognize others in order to fulfill what we lack in ourselves, with this lack and the care that it makes possible making us who we are. This would be a suitable beginning point for inter-cultural relations, for coming to some kind of peace both with the others we are held out toward as well as with ourselves.

AN ETHIC OF DIFFERENCE

This is the ontological framework underlying an ethic of difference, linking what it is to be some kind of being and what it is to be a good being, such that being (i.e., identity) and goodness (i.e., comportment toward difference) are related.[1] What distinguishes this ethical ontology is that identities are not considered as self-subsistent but are constituted rather in relation to difference. In ethical-ontological terms, we are individually how we are collectively with each other; that is, our being is essentially relational.

The thinking behind this can be extended beyond the strictly human sphere to include beings as such, with all beings constituted by their relations to other beings. If we cannot define a being by what it is not, we can trace out that liminal boundary between a being and its being in relation to . . . , describing the regions where a being is coterminous with what is around it. It is in these fragile shared spaces that the form of a being comes to presence, with a being showing itself precisely where it is in contact with the world. In this sense, if there is something self-subsistent to a being, it would consist in that being's relational capacities to be alongside of other beings.

So, to re-imagine a classical example of form and substance aligning, imagine the being of a tree, as defined in its relations to earth, air, sun and water. The defining boundary between the tree and its environment—that is, what is around it—is not so much a dividing line as it is a site of exchanges, with the roots of the tree being related to the soil, the stoma moving air in and out of the leaves, and the cell walls on the tops of leaves presenting a surface for the sun to shine on, and indeed in a sense also through. Similarly, in terms of the identities interior within the tree—that is, the various parts of the living being that can be named and distinguished—these take on their distinctive identities again in relation to the beings around them, so roots in relation to stem capillaries, stem in relation to leaves, and so on.

One could say that none of these beings would subsist in its identity without those essential relations to what is around, and not simply in the incidental sense of an exogenous system of inputs and outputs, but ontologically, with each being what it is only inasmuch as it bears out its identity in relation to what is around. Being is a set of relational capacities, contingent upon the fragile interrelations that bring beings into being, sustain them in their being, and move any one being on toward its particular ends. It is as if, within an amazing abundance and fullness of being, each and every being is propped up by those others that nestle against it, with each and every distinctive aspect of a being—every identity—defined by the inflections of a coterminous difference.

What is unique about human beings specifically, offering the opportunity for the project of self-knowledge and of a more ethical public life, is that these inflections of otherness into identity can be experienced in a critical, open, and caring way via a recognition that one's being is held out towards others or else closed off in an act of auto-mutilation that would try to cut away the differences that subsist within relations, instead mapping those now unmediated differences out onto others, like bloody masks of the self and grotesque parodies of identity. It is well to remember that this inward-turning cannibalism of the human spirit that would make monsters of our others empties out the self also, leaving an unrecognized nothing, a hollowness at the core, where those others and their differences belong.

Here we have a cue for inter-cultural discourse in what is a notably applicable principle—that identities are formed in relation to difference—that would initiate dialogue with something like the Socratic *elynchus*—that is the mutual recognition of a universal condition of longing. Out of longing, the search for self-knowledge, or what is the same, for a source of identity with others, is initiated by the recognition of an undetermined core of being that is the basis of free being. Considered within the context of contemporary human history, this would be somewhat of a departure. Tradition has come to mean assuming that in any inter-civilizational discussion the two or more sides come together and distinguish themselves, each with their own inherited identities and clearly formed ideas of what it is they want out of the exchange in the first place.[2] What is suggested here rather is that the proper beginning point for dialogue is the recognition by each of an otherness within themselves—a certain incompleteness that has drawn them into the mediated encounter with difference. Beyond this admission of "I know that I do not know" that Socrates insists upon as a necessary first step on the path to self-knowledge, in the case of inter-cultural discourse where peoples are thrown together, seemingly by chance contiguity, there is a mapping of this clearly registered lack onto these others; for as I do not know myself, so too do I not know these others.

This recognition of a gap within the self, imagined prior to the assignment of any distinguishing predicates to set apart self from self and I from I, is far from ignorant or empty. The freedom of undetermined being is not merely attached to a subject; it is at the center of the subject and sets the conditions of possibility for mapping distinguishing, extrinsic qualities onto a person. In this sense, one is free not because one belongs to a category of free beings—freedom can never be categorically assigned to another—one is free rather precisely because of the recognition of lack of such objective determinations.

It is this recognition of freedom as existing prior to the categorization and objective determination of human beings that is the ground of possibility for an identification with others. Even the category of human being becomes rather incidental in the exchange. Freedom thus conceived fulfills the need for an open signifier to reconcile the sort of determinations, borders, and inherited identities that naturally attach to worldly, temporal creatures, such as human beings. An original condition of freedom is necessary in order to make sense of those cultural differentiations within the recognition of difference, beyond assuming some sort of cosmological derivation (like, for example, peoples supposedly united by a common origin, as if they were born out of the rocks). From myths of origins to state of nature theory, no reductionist account of political life can satisfy the need of the human spirit for freedom or locate the basis of shared identity, just because identities so conceived represent enframing givens and therefore cannot serve as the grounds of a properly free form of association.

Even in the most everyday of exchanges with our neighbors, we can trust in the mechanical operations of weighing reciprocal actions—the *I'll give you this if you give me that*—only within the overarching context of some mediating third term. In the absence of a mythically ordered cosmos within which to negotiate our differences, and given the fact that different cultures have their distinctive myths and histories of self-interpretation, this opening at the center of every identity—the intellectual abyss at the core of every *I don't know*—makes way for paths to self-knowledge and the recognition of others in their differences.

The universal condition that is acknowledged within and that is recognized in the others who need us also is experienced as both necessity and freedom. It is the absence within that holds us out toward others in what are essentially needful relations—so I need my parents to bring me into being, I need my wife for companionship, I need my friends for free association—while at the same time those interiorized gaps are the bases for the self's own freedom, understood as internalized otherness. The self is free because it could be otherwise. Freedom is in this sense essentially undetermined being, that is, being that can become other than what it already is. This possibility of becoming holds the free self out toward its own otherness, moved in tremulous agitations between its being as it is and its being something else. The character of freedom thus consists in the quality of the relations between this being and that being otherwise: that is, is the self obligated to its given being, or does it exist in a free relation to difference? If relations are to hold onto that quality of freedom, they must begin with and return to the principle that our free being is basically undetermined—a principle that is also the basis of both critical questioning of parochial self-other identity formations and of the passage toward a more reflective and substantive recognition of difference.

This pattern of projected identities, whereby the self is seen in the light of its own otherness, can be employed consciously, to the end of seeing others in our own undetermined, hence free selves; or, alternatively the projected identity can take place unconsciously, such that we cling to given identities while sealing off this unsettling indeterminacy within, projecting the mysterious depth of free being onto apparently inscrutable others. Curiously then, when we imagine our barbarians it is they who turn out to be truly free, whereas our own identities are assumed to be fixed within a given natural order of things or historical legacy. So the barbarians become the wild ones—untamed, traipsing through nature—while we become the dutiful subjects of a civilizational imperative that is imprinted onto seemingly automatic fears and aspirations. How then can these priorities be reversed and new, more properly inclusive identities formed? This ethical question bears upon how to restore constitutive relations with others so as to see the other in the self and the self in the other, in what would be the beginning of the ontological conditions of care, rescued from a closure to difference.

IN DEFENSE OF LATE MODERN LIBERALISM

In the analysis of Jürgen Habermas, who is perhaps the most staunch intellectual defender of modern enlightenment ideals still standing, the twinning of identities via pathological reactions against a society's own absences sheds light on what is often treated as our contemporary barbarism—that is fundamentalism—a phenomenon that he maintains is of an essentially modern character. Fundamentalism consists of modernity's own shortcomings making themselves known. In this vein, Habermas observes that, "The enraged fundamentalist recourse to a form of faith that modernity has taught neither to become self-reflexive nor to distinguish between religion, secular knowledge, and politics, acquires a certain plausibility from the fact that it feeds on resources that the West seems to lack."[3] If the intellectual edifice of Western civilization is a construction based upon the separation of knowledges, then fundamentalism offers a whole worldview of reality reconnected to knowledge through faith. If the spirit of the West is subjectivized into lonely social atoms locked into a security state, then fundamentalism gives weight and collective substance to spiritual passions that may otherwise seem to drift past disembodied individual consciousness. This is not to say that fundamentalism is any kind of self-subsistent answer to questions of alienated subjectivity and projected freedom, since it offers no mediated synthesis but rather new forms of self-divisions, accompanied by a loss of the specific gravities of the various spheres of knowledge through their violent collapse. So, the transcendent value of religion is reduced to impassioned fervor, and political judgment conflated with divine revelation loses its footings in acknowledgment of the conditions of plurality that underlie every action. If fundamentalism reveals a drive to restore the wholeness of human experience by fantasizing about some retrieval of a past golden age—supposedly pre-existing colonialism and the self-divisions of spirit, most fundamentally modernization and modernity's division between subject and object—then the projection falls short of both reality and fantasy, unable to account for the manifold of differences that characterize worldly life and thus incapable of accommodating diversity in a global civil harmony. What remains rather is a railing against the hard-edged divisions within technological society, including its imperial conduits, and a projection back behind the history of modernity toward a recovery of a perfect autonomy experienced in a golden age of unconditioned origins.

As a symptom of the excesses of modernity, fundamentalism is revealing in its radical projections. Practically of course there is much that is troubling to it—cause perhaps to back cautiously away from the universalizing drive of modernity in its most singular formulations as irritant to fundamentalist discourses. Here one may find some indications of a renewed role for moderation—what has been a largely neglected virtue within modernity, like an ivy-covered statue to an important but half-forgotten god hidden in the back of some English garden. To such an end, and in what has become a characteristically hypermodern reflex, Habermas calls strategic attention to certain "cognitive dissonances": both at the heart of liberalism, wherein reason is employed not to determine a singular Good but critically rather as a mode of analytical insistence on the plurality of perspectives that is the basis of the political condition; and then also outwardly, as liberalism comes up against theocratic or totalitarian states and recognizes its own limits in adjusting their behaviors (as well as theirs in terms of exerting totalitarian control in a technological society). If freedom and equality together constitute the principles that move history forward, then a reasoned trust in these principles means keeping them pure of the contradictory impulse to force freedom. Here an admission to the existence of competing ideologies is offered as a strategic response to violent reactions against violent forms of modernization, and as a way to remain loyal to the commitment to diversity that gives liberalism its moral richness, born out of a certain exactitude of reason that seeks to situate even its own truths.

However, ultimately a commitment to pluralism is manifest as the true universality of liberalism—in an instance where an end is also its own means, acting as both principle of action and mode of accommodation. In this sense it is telling that in Habermas's critique, fundamentalists are ethically culpable precisely because they are not universal enough (e.g., they do not admit the apostates or pagans into their new order), and the proper, ethical response is, therefore, to be an even more rigorous demand for a properly inclusive world order, with an effective international law and global juridical order, a standing army capable of global policing and under international sovereignty, ultimately with its basis in a self-consciously global citizenry. Over and against a false, almost farcical universalism (one that cannot bear a world with Buddhas in it—an impossible utopia if there ever was one), Habermas calls his readers toward a "*strict* universalism,"[4] as if to accommodate the fundamentalist reactions against modernity by radicalizing and clarifying their fundamentalism more fully than the fundamentalists themselves are capable. If there is a tendency for the modern spirit to exhaust itself in anxious relativism, then perhaps one can take a cue from the vehemence of fundamentalist outlooks, so as to make the modernizers more fundamentalist in their beliefs, while the fundamentalists (it is hoped) are made more properly modern. This would be to effectively flip Yeats's rather grim assessment of the spirit of modernity, wherein:

> The best lack all conviction, while the worst
> Are full of passionate intensity.[5]

A similar note of radicalization from within the modern spirit can be heard in Slavoj Zizek's response to the neo-liberal corporate consensus, toward certain projections beyond its dominant logic. He offers a favorable comparison of his hoped-for intellectual vanguard with fundamentalist Christian soldiers, specifically endorsing their unblinking sense of rightness—of divined access to the Truth—alongside of a rejection of lazy pragmatism as no path to virtue, whatever technical successes that default position may offer. Rather, the virtuous position must know itself as virtuous and act that way. Here a commitment to one's own essential rightness is put forward in order to short-circuit the smug but empty belief that equality trumps virtue, a liberal tenet according to which ethical goodness is indexed by the disavowal of any such standard.

In contrast, the authentically fundamentalist position for Zizek would be something like the following:

> In short, goodness (applying common moral norms) which is not directly grounded in the Gospel is ultimately just a perfidious semblance of itself, its own tragedy. . . . The basic premise of this book is that, cruel as this position may sound, if one is to break the liberal-democratic hegemony and resuscitate an authentic radical position, one has to endorse its materialist version.[6]

What comes to the surface here is both the radicalism inherent in fundamentalist social movements, as well as that identity's mirror-opposite truth. The strategic concern of both is with the remarkable capacity of liberalism to co-opt divergent positions while eviscerating them of their ethical substance, whether that substance be of cultural or historical origins. If one does not buy the liberal proposition that any essential political debates are settled by the groundwork proposition that are all equal before the law and chooses to hold out for a divine law or for a more substantive, revolutionary social justice, then some non-negotiable demands must be made, inspired by a sense of the perfect rightness that is encoded in their impossibility.

And beyond strategy, isn't this the real basis of refusal to negotiate with radical groups, because to do so would introduce certain non-negotiables into a political process that rests upon the leveling of political positions to a homogenous plurality, a starting point of incidental differences that is only resolved toward consensus when the various interests involved recognize and internalize the contingency of their interests, that is the relative value of their own positions? There would thus appear to be no room for essentialism in liberal politics. Even the inherent value of individuals and of groups as essentially free and equal is largely a hypothetical principle, as in let us act as if that were the case (while recognizing that it is more often not). For those demanding more than a hypothetical equality, asking—*Where's the divine measure?* or *Where's the revolution?* the ethical weight of their demands are indexed rather by their degree of impossibility. So what is corporate management to do when the new barbarians demand the right to piss in different colors? According to the "authentically radical" position this seems to be the point, that there is then nothing that can be done, no tit for tat accommodations possible, no corporate sale or exchange, and an impasse arrived at as the barricades go up. And the hope then is that by the pulsed interruptions to the seemingly inexorable, osmotic flows of opinions, resources, and capital, certain essential gaps—new political margins—will be opened up to reveal real difference, and difference not understandable *qua* difference, that is, as mere novelty in the absence of particularizing context. What it is and what it will be then become active political questions restored to their full ontological dimensions. There is action again, and uncertainty, though this paradoxically requires a determined certainty that the impossible reassert itself as the locus of change and the true basis of political reality.

In this spirit and with all the wild fervor that one would expect of a self-proclaimed vanguard Zizek insists that:

> . . . the "Real as impossible" means here that THE IMPOSSIBLE DOES HAPPEN, that "miracles" like Love (or political revolution: "in some respects, a revolution is a miracle," Lenin said in 1921) DO occur. From "impossible TO happen" we thus pass to "the impossible HAPPENS"- this, and not the structural obstacles forever deferring the final resolution, is the most difficult thing to accept.[7]

Across ideological lines, we hear a very similar emphasis on the political importance of the impossible in the altogether more cautious voice of Hannah Arendt, in her account of the unpredictable nature of political reality. Arendt would disagree with Zizek's and Lenin's identifications of violent revolutions as miracles, considering the all-too-mundane element of apparent necessity that makes the kind of revolutions that Lenin had in mind irresistible, acting more like the unstoppable revolution of the planets reproduced in human affairs than a true transformation of the conditions of human existence. Rather, action in its fullest sense must have an element of the impossible to it, projecting out beyond the circle of conditioned behaviors toward futures that even the actors themselves cannot predict. Action after all requires the presence of others, in whose eyes one can never be entirely sure how one's words and deeds will be interpreted.

> Hence it is not in the least superstitious, it is even a counsel of realism, to look for the unforeseen and unpredictable, to be prepared for and to expect "miracles" in the political realm. And the more heavily the scales are weighted in favour of disaster, the more miraculous will the deed done in freedom appear; for it is disaster, not salvation, which always happens automatically and therefore must always appear to be irresistible.[8]

In terms of where one could go looking for such miracles, I would suggest that they have less to do with violent revolutions (for the reasons that Arendt identifies in describing the seemingly unstoppable character of revolutions once in motion[9]) and pertain more properly to acts of inter-cultural recognition and translation. For as far as politically significant determinations go, the logics and lexicons embedded in languages and the practices and perceptions transmitted within a particular culture are the basic conditioning effects that shape our being with others. How then is it possible to move from one grammar and logic and one particular shape of reason to another? Inter-civilizational dialogue of significance is not reducible to syllogisms or mathematical formula, so how can we know what others are speaking toward in the absence of common words inbetween? To do so requires: first, a recognition of the other as a being like oneself; and second, a translation of one's own desires into a third term that overarches the particular perspectives of "we" and "they." Traditionally, at least within the Western tradition to its earliest written records, these mediations have been negotiated by assuming the existence of a larger cosmological order within which our fractured political identities are situated. Here the miracle is invoked in a transcendental gesture that lifts eyes up from the mundane plane—and the particular thing before my eyes that both you and I want (say for example an apple, of which I would prefer both the first clean bite and to nibble around the core)—to an all-inclusive perspective, from which one may rightly ask, *What does it matter if I or they eat the apple? Or if they get the first bite and I get the core?* This kind of transcendental miracle is not ultimately about economic sharing, which is more commonly negotiated in terms of reciprocal actions, as in Aristotle's theory of distributive right,[10] a pattern of thought that can be traced all the way back to Homer's heroic society, where honors are indicated by the place at which one sits at the common table, and the cuts of meat and war prizes distributed to each of the heroes according to their actions.[11] Rather, the miracle of a greater good makes all of the tit for tat of reciprocal actions, conspicuous consumption, and the manifest differences that those customs imply seem rather insubstantial. Thus Plato can deride the Homeric heroes for being selfish because they fought in expectation of manifest honors in return.[12] When a transcendent Good is invoked as the mediating term that connects the interests of all with all what is hoped for rather is a release from the driving determinations of personal interests. In its perfect instantiation, the quarrel over food at the dinner table is cured by a polite self-denial—a *you first/no, you first* that is only finally reconciled by what are themselves incidental habits of politeness—for example, the serving bowl goes around the table from left to right. Yet if the model were carried to its miraculous extreme, no one would eat at all; in the face of this most extreme of miraculous interventions, detached from the vital habits of self-interest, perfect unity is thus projected toward a kind of absolute peace (like death), where individuals find their union with the cosmos, though at the cost of their own desires.

At a practical level, the reconciliation of differences through transcendental means requires some sort of self-denial that leaves room for others to make their own self-denials—that is, once they have eaten their fill and realized that life is not sustained by apples alone, then they can go on, as we have, to imagine and then recognize an other beyond the I. Properly absolved of particular desires, self and other then enter into a union within some cipher of their projected, common desire—the thing that we all want and that is so large as to encompass us all instead of us encompassing it (say, as one takes in and digests an apple). This would be the perfect, indissoluble form that subsists irrespective of our particular attachments to it—a kind of cosmological apple that we could all take first bites off of forever, though only once we have each and all transcended our particular desires for apples as such.

This tried pattern of projected, de-linked desire would certainly qualify as a miracle in the sense that either of Zizek or Arendt use the word. And it is this kind of political miracle that is so often invoked when mediators in conflict situations implore antagonists to set aside their differences—by which is meant their fixated, personal forms of desire, that is, what each wants for themselves—in favor of a more universal form of longing, like peace. The question is how to fill in that ultimate desire beyond the negating terms of a *lack of unfulfilled desire*, which is its usual formulation (not a situation that can be expected in political reality, where enlightened saints are the exception).

Since we are entering into the domain of ciphers, with symbolism standing in for an object of desire that by definition cannot be singled out from its environment if it is to remain truly overarching (short of delineating a philosophical definition of precisely what is this objective fulfillment of ultimate desire in and of itself), it would help to have a story, the narrative arc of which points those who hear it toward a common projection. In this we essentialize our shortcomings by telling a story in reference to them; thus, to take a familiar example, the apple is elevated from an everyday object of desire—something that all of us have, at some time, likely had a hankering for—to a symbol standing in for self-knowledge in the Biblical story. Now this symbolic object—"the fruit of the tree"—is taken for its universal significance: It is what each of us knows of ourselves—the singular *no* that distinguishes each one of our identities through a negation of the environment as given, setting apart each human being in their freedom and finitude. It is the knowledge of nature, human and non-human, with awareness of reality, including the foreknowledge of one's own death, insisted upon. It need not even be an apple: A pear may have been more historically accurate, but whatever the thing itself may have been, it is the symbolism that counts.

The obvious difficulty with this mode of reconciliation via symbolism is that it requires, if not an immediate intuition of the value of the symbol (there is no natural reason to suppose an intrinsic link between either apples or pears and self-knowledge) then rather an overarching narrative order within which the symbol can be situated. And as inter-cultural relations go, one begins precisely with differences between various founding myths. And while it is true that certain symbols have resonances across various traditions—as in, for example, stories about floods, or gods that die, or the scapegoat—none of these symbols carries an immediately intuited symbolic value along with it. Rather, all mythic symbols are characteristically open in what they could be taken to mean. If this weren't the case, then the symbol could hardly function as a cue to raise our sights—apples would mean apples and floods would mean high water levels, hardly of spiritual significance in themselves. Inasmuch as they function as symbols, a huge range of interpretations is admitted to these universal symbols.[13]

Here something like Joseph Campbell's popular encyclopedia of resonant mythic symbols and his efforts to demonstrate their cross-cultural likenesses must have the unintended effect of de-mythologizing the stories he relates, just by assigning universal, rationalized meanings to them. Ultimately in his telling, and in the broad tide of history, all peoples are drawn by the same currents, and all social manifestations of deities become as many masks of God, moving toward the acme of faith, considered along the lines of the God within you.[14] From Homer's *Iliad* to George Lucas's *Star Wars*, from this view religious sensibilities are developed at a cultural level through the retelling of stories we already know. Rather, I would argue that symbols gain their significance not from what we already know about them but from what we do not understand. If there is a basis for universal identity, it is this: What we have in common is what we lack in ourselves. Or put otherwise: contra the desire to converge the world's myths into an amassed singularity, reducing the projects of self-knowledge and inter-cultural understanding to the self-projection of an assumed identity, identity is not sameness; identities consist rather in difference.

BARBARISM IN ITS CULTURAL SPECIFICITY

What this can be taken to mean is that the halfway ignorant, halfway illuminating taxonomies by which barbarians are classified are in fact quite culturally specific, with what is a barbarian shifting radically from civilization to civilization. It is apparent that two civilizations could regard one another as barbaric, with each defining itself in opposition to the other; however, I would argue that the alterity of barbarism goes beyond a simple rejection of the ordering principles, worldviews and practices of another people. Rather, there is a specific content to the category of barbarism, even if it has little to do with the people who are put within the category themselves. So what was barbaric for an Athenian of the Classical Enlightenment, for example, would be distinct from the modern European Enlightenment's imagined Other: The ancient Athenian barbarian was tough and manly against what was perceived to be an increasingly feminized Hellenic culture, exemplified in leisurely, coiffed Athens,[15] whereas the European's barbarian was often represented in quite opposite ways, as a feminized, passionate Other, coquettishly hiding from the light of scientific reason; the Greek barbarian was a no-nonsense, straightforward talker (Anacharsis, champion of the plain and obvious), whereas the barbarians that Napoleon sought to liberate from themselves were seen as creatures of subterfuge and mysticism.[16] In each instantiation of supposedly radical otherness, there is a glimpse of certain capacities abandoned by a civilization in its increasingly artificial orders. If only then that these others could be reconciled with, one might continue along the line of thought, so that the too-civilized, artificial society could be made whole again, those vestigial capacities reconnected to and suppressed energies of self-overcoming re-invigorated in a civilization that sees itself as having lost its roots in reality. This is the impetus of self/other formation that, I would argue, is at the core of fascination with our barbaric others.

ON THE IDENTITY: BARBARISM = NATURE

The conceptual flexibility at the core of notions of barbarism bears a striking likeness, perhaps not surprising, to the variance across cultures in ideas of human and non-human nature, in that both the natural and barbaric are projected as foils for a society's self-understanding: Imagined others representing exteriorized gaps in identity. By this inversion of identifications, where a self knows itself as if in photo-negative form in its imagined other, a perfectly natural and fully self-conscious society would then be one with no conception of nature, nor of itself as constituting "a people." These blissful ones, we can imagine, would feel no need to go about constructing a nature to wrap themselves in; nature would already be busy at work in constructing them instead. Or at least we can imagine something like this, and by so imagining bring nature into a social mode of being.

Another way of imagining nature is as that which will always be. This sort of imagined disposition has been devolved into a bare sense of the apparent givenness of nature by the leveling effects of modernity; a process that had much of its success precisely by reducing nature to a set of resources, that is simply as "stuff"—objectively determined and determinable realities thrown before humankind. This is nature due to be integrated: being that which just happens to be, hence having no intrinsic purpose of its own.

This has been a key liberating idea for modern science, which had its inflation of value when teleological physics was abandoned and purpose liberated to human historical action, considered then to be cumulative over the course of processed time, as in a technological form of progress.

SENTIENCE OF NATURE

One of the defining characteristics of the prevailing, current mode of conceptualizing nature is that the concept formation assumes that nature can be quit, that there is a stepping out of nature that is somehow possible, as one steps out of an apparently given, externalized nature and into a technological order of things.

In this regard, in critically stepping back from these dual phenomena of technological leveling coupled with a reactionary politics of naturalized self-other identity formations, the patience of other animals is perhaps instructive: that they are patient because they have no other place to get to, no environment as such to flee from.

It has become a matter of academic consensus to say that our notions of nature are social constructions, that there is no self-evident "nature as such," but that our frameworks for understanding natural phenomena (the subject reduced already to the philosopher's formal schemata of intellectualized experiences) intervene and mediate between self and world. Informed by such an awareness of how such dichotomies have become entrenched in thinking, Joel Kovel writes that,

> Nature palpably exists irrespective of what we say about it. And yet nature only exists *for us* insofar as we say anything about it. All propositions about nature, from the most esoteric investigations into cosmology, to the regulations for dumping waste, to the writings of ideologues left, right and center—including, to be sure, the thoughts written down here—are mediated by language, which, besides being an imperfect mirror of reality, is densely social and historical.[17]

While it may be true that our conceptions of nature—the frameworks within which we collect the stray bits of sensation—far from being abstract, universal totalities mapped onto the world, are themselves situated within a particular cultural context, growing out of a people's history like saplings out of cracks in fractured bedrock still one ought to ask, how do these variable ideas of nature arise in the first place? From whence the insistency of nature? Where conceptual apparatuses fall short of the thing itself, I would suggest that it is in the lacuna of our socio-historical revisions of nature to suit our seeing that we may find clues as to a presence before which our schemata seem like so many empty categories waiting to be filled in. For isn't even a hermetically sealed, *verum factum* epistemology itself made possible by some presence of the other in consciousness? Do we not find some flecks of pollen and stray wind-blown leaves inside of the doors of perception?

In the spirit of a similarly imaginative reason, and in gentle mockery of Kant's theory of perception, (namely of a phenomenal world wherein the mind makes the object) Luce Irigaray writes:

> But now everything has moved inside the house the subject has made, or is. And whether the scene seems set inside, or outside, whether in his room or in his study, sometimes enjoying a fire fancied to be burning in baroque curls of smoke or else gazing out through the/his window at the still in(de)finite space of the universe, the action is always inside his house, his mind. And what or who can now put it outside? Only a messenger of revolution perhaps? Or else the fact that this hearth is made of glass and that those glasses—rather tarnished by age, their brilliance dimmed, having always in fact been unsilvered or blackened by smoke—mirror so deadly a boredom that, whatever one's firm intent, one might finish by wishing to die—to die of love, were that still possible—rather than have things just go on. Forever.[18]

The "Forever" implied in such a self-made perception by categorization is a process, one can imagine, somewhat like decision-making by committee, where even if the results are not true, the foibles are transparent—a paradoxical but predictable extension of enlightenment logic to its own inherent limits in the irrationalities produced by reason.

As Irigaray hints, all sorts of non-transparent senses—from boredom to love—can play their roles here as potential causes of a revolutionary return to nature, which even if nature cannot be pinned down as the cause of sensation, still unsettle the quiet mind enough to go outside and muck about in a reality that is wonderful because it is both unknowable in itself and behind all that is known.

Philosophers of perception have long wondered over the apparent intentionality of phenomena and how this sense arises. Keeping within a social constructivist framework, one may ask whether this intentionality or the apparent "out-there-ness" of phenomena is itself a kind of special effect of the synthesizing imagination, a projection traceable to origins in the structures of perception. In this train of thought, as Hobbes rather curiously suggests, the perceiving subject registers the objective insistence of sense events via a certain "counter-pressure" delivered from the mind (he is not sure on this himself, and writes that it could be either the mind or the heart that does the pushing back), by which the self seeks to "deliver it self" outward into the world—claiming these sensations and the world of nature as its own.[19] By his account, the world can only *seem* to maintain an objective quality via a projected objectivity of sensation.

This inherent subjectivism is what requires Hobbes's blatantly propagandistic political methodology—for what if not all subjects fear the objective, natural fact of violent death? His answer is characteristically self-referential: that they then can simply read his book and "read thyselves," or alternatively, suffer execution as vainglorious martyrs—with either option being fine for a state conceived as machine for the production of social order, whatever reactions to the management of the self and the social there may be notwithstanding.

The pre-linguistic barbarians of undifferentiated sense experience that bump and careen their dangerous ways through Hobbes's state of nature are of course nothing other than imaginative projections of particular qualities abstracted from the social (dis)order that Hobbes found himself in: that is, an imagined nation of shopkeepers motivated by bourgeois, self-interested sentiment minus (and this is Hobbes's hope) religious and populist extremisms. What is of significance within his projection of the primitive and others like it, I would argue, is what is left out of the imagined natural state, that is, what is considered to be unnatural in ourselves and our societies. The only real essence that is hinted at here is thus the unspoken one: that aspect of ourselves that we would wish to occlude via an identification with certain inherent, self-subsisting qualities that exist somehow "out-there" in our externalized history and pre-social origins. It is a rationalized myth, to be sure, but a telling one that needs to be interpreted toward an intuition of the limits of our artificial social orders. So while we may not come to know what our own nature is in itself, short of jumping over our own self-made shadows,[20] what may be aroused is an awareness of certain urgently felt absences that exist within the orders of our societies and ourselves, coming up against the limits of our capacities to bring order to the world and finding a measure of self-knowledge in that modest acknowledgment.

Other contractualists are clearer even than Hobbes on the rhetorical and sociallyspecific dimensions of such attempts to project back before the distinction between subject and object, toward a world of subjects determined by objective impulses. Rousseau for one acknowledges that there can be no rationally based knowledge of such a state, precisely because any properly natural awareness would have to be pre-linguistic and pre-logical—a kind of amorphous soup of drives and aversions. To the extent that one wishes to describe something revealing about the state of nature and those whom inhabited it, some one element must be singled out, with "Man" thus standing back from his previously unsymbolized environment—hence apparently outside of original "Nature." Yet on what basis could such an element be called out from an unstructured field of relations, within which, according to Rousseau even species distinctions are absolved by a perfect unity? There is good reason why Rousseau, includes a modest "I know that I do not know" at the outset of his *Social Contract*—and it is not because of any old-fashioned sense of virtuous modesty. Rather, the degree of unity that subsists in his original state is so compact, the love so perfectly self-realized, that it is difficult to imagine any reason why it should be left behind. For what would incline the fully adjusted being to set up the first fence, when all is his or hers to begin with? So Rousseau can only suggest that self-preservation and sympathy, as the basic human motivations, are hypothetical or experimental abstractions from an original totality, but just how are these hypothetical traits formulated?[21] If these are our first impulses, what intuition calls them to attention as distinct modes of a hitherto whole consciousness?

Rousseau's critique of other social contract theorists like Hobbes and Locke, intellectual supporters of the notion of the naturalness of property rights, is that they "have transferred to the state of nature ideas picked up in the bosom of society."[22] It is apparent that the mirror-opposite truth can be attached to his reflections: that he imagines certain unfulfilled desires—say for social solidarity, for a life in harmony with nature, and for love without limits—projected back onto humankind's earliest origins. Clearly there is something circular to this pattern of thought; for isn't the purpose of his prehistorical narrative to provide a standard by which to know and to judge the deprivations of the present, specifically in the case of Rousseau, inequality among men? It appears that there would need to be some middle term to connect and separate the artificial social order of the present to and from the natural state of being out of which our current conditions arose. Such a way in between is made that much more inscrutable by Rousseau's insistence, now commonplace within social constructivist schools of thought, that reason itself is shaped by the social and historical context within which it operates.[23]

Actually, this skepticism toward the transcendental value of reason is fairly consistent within modernity, stemming from an original inversion of the faculties that puts reason in the service of the passions. From Hobbes's characterization of reason as the "scouts and spies" of desire, allowing us to get what we want without informing us as to what we ought to want,[24] to Marx's treatment of reason as ideology informed by historical situation,[25] to the Frankfurt School's indictment against instrumental reason and the systemic ways in which it shapes human being in the world and with others,[26] reason does not maintain the high ground that it had in classical philosophy, or even in medieval thought, however much the Schoolmen may have vexed over the question of the proper relation between faith and reason. Reason comes to appear rather as a rule-based synthesizing of experience, as a process, a way of connecting this phenomenon to that in thought and practice. The working out of that particular version of the imaginative impulse that is reason leaves nothing untouched, not even its own conditions of possibility and protocols for reorganization; that reason consists, therefore, in recurrent struggle to see the mind's order of things in reality. Acknowledging this gap, and the self-subsistent tension between imaginative reason and the unsymbolized real, or self and other, that result, is the first crucial step toward critical self-knowledge in an otherwise self-rationalizing society.

ON THE DELPHIC ORACLE

> Love of one is a barbarism; for it is exercised at the expense of all others. The love of God, too.[27]

What Nietzsche sees as the barbarism of monotheistic religion hinges upon a kind of exclusion that disallows the piety of others by laying claim to a god. The specific pattern of exclusion works by the act of laying claim itself, as if one were to know the god as an extension of the self, in what is a particular frailty of monotheism under conditions of modernity. In a liberal humanist, god-made-man spirit of late modernity, knowledge works by proxy in a *verum factum* code, that is, to know the world as we have made it. Very much within the spirit of late modernity, the fundamentalist lays claim to a god, whereby their exclusion of others is a natural consequence of a political ontology where identities are considered in antithesis to an irreconcilable and unknowable other, as in for example a religious directive to attempt eradication of statuary to the Buddha.

In contrast to the overweening character of contemporary monotheisms, the classical Greek tradition evinces a deep awareness of the impasse between the thinking subject and its tendency toward objective determinations—that toward the transposition of an order from within the thinking self onto the world at large. So although the notion of a parity between the workings of human mind and the order of the cosmos is assumed by most traditions that share in some kind of cosmology, far from revealing the order of nature as self-evidently obvious, such intuitions that the world is somehow as we understand it deepens the mystery of phenomenal reality by translating the glowing, shifting field of outward appearances into subjective, connecting events of consciousness.

One of the inscriptions over the entrance to the temple of Apollo at Delphi is the well-remembered "Know thyself," an imperative that Socrates would abstract from its circumstances and immortalize as his personal dictum; though tellingly, his quest for self-knowledge was at least partly inspired by a desire to prove the Oracle wrong. Socratic self-knowledge begins with the no's one says to oneself and ends with their refutation in what Hegel would call a negation of a negation, which is also a kind of confirmation. So Socrates refutes that he is the wisest man in Athens and, by the strength of his refutation, proves that he is.

Through all of this, do we as readers come to feel any closer to the person himself? Hardly at all, as it seems that the man behind the questions slips away into a stream of *neither this, nor this, nor that* . . . . He is the deepest thinker because any question that is asked of him in return, like a stone thrown into a bottomless well, finds no echo. Epistemologically and ontologically this endless course of questions, and of questions to questions, is entirely appropriate to one who claimed expert knowledge exclusively in erotic longing, that is consciousness of the condition of essential absence that moves thinkers to think and lovers to love.[28]

For those who claimed to know who they were and what they wanted and reciprocally demanded this assured identity of others too, this was a source of regular exasperation. So a characteristically indiscrete and transparent Alcibiades complains in the *Symposium* that there is no consummation with Socrates, no getting to really know the man himself. Adapting his teacher's notion that the real being of a thing is wrapped up in its likeness—that is, what a thing is like, or in Plato's language, its *eidos,* literally the look of a thing—Alcibiades, who is himself elegantly attuned to looks, compares Socrates' outward appearance to a carved wooden likeness of a silenus—half-human, half-animal—within which, as if behind outward opening shutters, are contained "images of gods within."[29] What gods specifically we as readers are not told, but in any case we can register in this revealing metaphor that there outwardly appeared to be something hollow to the man himself, who is here filled in by the images of place-keeping gods, which, *qua* images, refer beyond themselves. This capacity for inward projection would fascinate and, in the case of Alcibiades, frustrate the erotically charged student of philosophy, while holding the philosopher out toward ciphers of desire freed from objective determinations.

Reflecting on the inward thinking life of Socrates, what these metaphors of the hollow man and the gods inside of him suggest is that self-knowledge is not cumulative, as in an ongoing attachment of various predicates to an assumed solid core of a human being—something like a personal page on a social networking site, where a person describes themselves by attributing various qualities, status, interests and occupations to their avatar—but rather, there is an opening at the very core of a human being into which one could pour any number of predicates without filling the self in. One does not sum up one's personhood by identifying as a poet, as a politician or a carpenter—to use the particular professions that Socrates interrogates in his *Apology*. What these assigned functions indicate rather are the limits of the self as it is circumscribed by a specific technical capacity. And yet even still, within these bounds, if instead of accumulating functions we strip away these predicates to find the naked self underneath, what remains behind the masks is a certain nothing—the abyss that one witnesses staring into the perfectly black pupil of another's eye. For beyond the brown or blue or green iris there is this colorless wonder within that amazes to the extent that it is freed of attachments to defining qualities. This is a self that cannot be reduced to information because it has no given, determinate form, and is hence unknowable in any reductive way, existing instead as an essentially free being.

Beyond Socrates' exemplary know-nothing form of self-knowledge, embodied in the "Know thyself" that each self is, it is worth recalling also that there were other inscriptions along the way to that fissure in the rock at Delphi, from which underground fumes apparently acted as inspirations delivered by the god of light and reason to send young, female oracles (as a rule and custom also poor and often sent unwillingly) into convulsive spurts of words.[30] Carved in stone along with the Socratic imperative to know oneself was its mirror injunction: *meden agan*, or "guard against excess." But how would one know if they had gone too far, if the drive is for self-knowledge itself? Or these two aphorisms—"know thyself" and "guard against excess"—could be taken as equivalent statements, one in the positive and one in the negative, with self-knowledge equivalent to a cautious recognition of one's limits. Thus the being of the self would be defined in relation to its non-being. This sets up two impossibilities. First, if we understand the defining boundary or, in Platonic language, the form of a thing in relation to its non-being, with essence circumscribed by environment, still we cannot come to any conclusion about the essence of a thing though a register of what it is not—that would be an endless game of *not this . . . not this . . . not this. . . .* And second, if self-knowledge is dependent upon knowledge of the self's otherness—in the most basic terms the world as other—then we can come to no surer sense of self than we can pin down the world of sensation. So we are left knowing the world as transposition of reason and knowing reason as a reflection of the world. What circularity presents itself here is far from an insular one, taking in the world as whole and self as microcosmos of that projected mystery.

There is a third, saving element included over the portico before the cave at Delphi, reading simply "You are."[31] Here we have something like Descartes's *cogito ergo sum,* though in a highly compact, and thus inherently complicated form. Whereas Descartes's *cogito* stakes its claim on a supposedly self-evident proposition that attaches being to thinking to the I that thinks, the Delphic aphorism begins with being and leaves whatever you may or may not be to reality and the imagination. If we cannot summarily in a few words know either ourselves or the world, we can at least intuit this from the identity of another, communicated in the form of a "you are . . .": that one is, that in general beings are and specifically human beings are. Though it is nice to have that inkling, the saying falls tantalizingly short of attaching any distinguishing predicates to the being beyond or the human being within. And yet it is not simply a matter of intoning *Being*, since the arrested phrase does point out beyond itself, ascribing being to an-other being, as in the "You" who reads it. What we are not reassured of is anything so definite on the part of the author (who vanishes like any good graffiti artist).

As a kind of pre-emptive anticipation of the post-modern reaction to Cartesian self-certainty, I think that the Delphic aphorism in many ways answers Nietzsche's thorough-going critique of Decartes' *cogito*, by relocating the quest for self-knowledge from an insular reflection on the operations of mind toward an awareness of the constitutive projections of alterior consciousnesses onto others, through which we achieve a glancing self-awareness in the acknowledgment of our self-cast shadows. Well beyond being a clear and manifest proposition, Nietzsche suggests that the *cogito* should be considered as more of an aphorism itself; that is, its truth is its hiddenness. In reality, far from being transparently obvious, the *cogito* functions as a complex amalgam of assumptions and anticipations: beginning with the most curious of immediately given of "things in themselves"—that is, thinking—an activity assumed to gather together a multiplicity of operations of mind and subconscious drives; then, as the syllogism/aphorism unfolds, ascribing that activity to the I that thinks (alternatively, couldn't we say that it is the I that is thought?); then by consequence settling upon thinking (and specifically thinking of the clear and perspicuous variety, ideally reducible in its propositions to syllogisms) as the defining actions of the self; and further, assuming that as an action thinking must have some sort of cause (again preferably singular) thus carting in a naive conception of causality as another apparently immediate given. This is quite a lot to claim to know without having opened the doors or windows of the mind. Yet short of reducing the syllogism from a simple truth to an equally simple non-truth, Nietzsche leaves off his critique with a smiling flourish of charming subterfuge and (grammatical and logical horror upon horror!) a doublc-ncgativc:

> "My dear sir," the philosopher will most likely give him to understand, "it is in truth unlikely that you are not in error—but why must we have truth at all cost, anyway?"[32]

The sentiment would not be lost on one of Apollo's priests or oracles. One rambunctious philosopher of the "clear and distinct truth" could put them out of their business (unhappily for the priests who profited, though perhaps liberating for the young women compelled to a life of religious service by poverty and drugged into obscure thinking by the god of light and reason). For what need is there of an externalized, Apollonian god of reason when the *logos* is reduced to grammar school logic lessons, or of oracular advice when reality is rendered as the law-bound operations of a mechanical subjectivity—the mind as a thinking machine producing objects? Part of the lesson here is that immediate givens ought to be left to the subtle contingencies of judgment and hints of intuition, whereas reason is more properly deployed to demonstrate the inexorable complexities of constitutive relations, hence serving as indicator of its own limits *qua* reason. When a very few words are treated as crucibles of understanding, they can be *intuited* to mean something perfectly simple (this is after all the active mystery that keeps one thinking after them), while the same words are at the same time *reasoned out* to imply a whole world of potential meanings, including those other than our own.

The inscription over the portico at Delphi was taken as a bit of a mystery even at the time. A conventional line of interpretation was to fill in the apparent ellipsis with a "nothing," as in "You are nothing," in what would be a proper token of admitted modesty for a pious devotee to Apollo. And yet even if the saying is projected in that direction, the presence remains intact, such that the open absence of the "you" takes on a substantial sense. After all, presuming the "you" refers to a human who can read the words, there is a lot that can be filled in this way or that with a human being. The words can be interpreted toward various meanings, much as a person can take on a whole host of purposes as their own. "You are". . . a perfect aphorism for reflection that shelters the mystery of self-knowledge behind seemingly obvious truth.

PROFANING THE ELEUSINIAN MYSTERIES

> "At least I'm not innocent! Nobody can say that!"[33]

The charges brought against Alcibiades were said to have included impiety as well as treason. The most serious of the charges of impiety relate to the profaning of the Eleusinian Mysteries, what were the most sacred and solemn of the Hellenic religious rites, and most intriguing even still to this day, precisely because they have remained mysteries. We have outward accounts of the stories that inspired the rites, for example, in the Homeric *Hymn to Demeter*, but since this was a matter of common knowledge at the time one cannot identify this text directly with the Mysteries themselves—that is, then and now, from a hermeneutical perspective, the Mysteries meant something more than their representations.

Beyond their simply being mysteries, there are, as in any really well-kept secret, some tantalizing truths to be explored in what references to the Mysteries remain. It is said that of all the cultural rites disseminated by the Athenian empire, this one was unique in fostering a sense of unity and concord among those who participated in it.[34] This says quite a lot, if we consider some other fixtures of that so very social Hellenic civilization: the Olympics with its festive admiration for human bodies in their finest form; or the spread of democratic forms of governance which located political authority in the masses of humanity; or even the orgiastic bliss experienced by the revelers of Bacchus—that feeling of oneness in a mob. Surpassing all of these in sympathetic sociability, the Mysteries were said to have fostered a sense of organic wholeness, something like the circular perfection of the seasons—of Nature's ascendancy into spring, its quieting in autumn, death in winter, and rebirth again—except experienced by a group of people in microcosm, though somewhere behind walls and unreported.

The secrecy of the rites was taken so seriously that any who put them out in public was punishable by death and disenfranchisement—an erasure of the person from personhood that would return their being to the same darkness that they had inopportunely shone a light on—a punishment enacted in binary conjunction with the infamy of a public record of the crime, inscribed in stone and put on display in a public place for the shame of the dead. They would thus be made into nothings, voids whose only public record would be their own executions.

Of course, these ancient prohibitions did not and do not keep one from wondering what the initiates into the Mysteries came to do, know, and feel. And yet, to make an historical distinction, one would imagine that this desire for the "thing itself"—that is, the rites and secrets communicated inside of the temple walls—would manifest itself differently in different times, because in each historical instance the focus of desirous attention would be delimited by the specific absences present in any one particular cultural milieu.

In one of the rare passages when the Abbé de Barthelemi inserts himself into his compendium of Hellenic culture, he attempts to get behind the narrative account of the initiatory stages that screen the rites from public view, apparently as told to Anacharsis the Younger by an initiate. So the author attaches a note accompanying the pious initiate's "we saw and heard things which it is not permitted to reveal,"[35] wherein the scholar of antiquity notes archeological evidence given at the supposed site of Eleusis, where the Mysteries were enacted, and sketches out a likely architectural layout of the temple, which he describes in a proper Enlightenment (if reductionist) spirit as having been much like a cinematic amusement park. Therein, we imagine visitors herded from one room where they are shocked into wonder by the appearances of shades and haunting voices, to another where they witness an enactment of Tartarus, and last to a sacred enclosure where statuary to the hidden gods are kept. In the Abbe's interpretation the Mysteries are sought in their physical ground, with all the scrupulous literalism of an archeologist who would go looking for a people in their pottery shards. To be sure, what remains *is* one index of reality in its subsistence; however, it is easy to imagine that others might go looking by different ontological criteria, and by different evidences. Rather than the rock-solid artifact, another investigator might go looking for the Mysteries' subsistence in their emotional resonances, say in the dramatic arts. So we might dig into Sophocles for a clue as to what is hidden, and hear Oedipus lament. . .

Why should I see, when sight showed me nothing sweet?[36]

Or the historically minded might put their trust in Thucydides, that most impartial of the Athenians—a general able to imagine himself in Spartan sandals. The historian relates that Alcibiades's crime of profaning the Mysteries is traceable to an event, a particular night even, when a group of drunken friends made mockeries of the processional rites about to begin, with the drunken *komos* (root of comedy) carousing through town as if they were priests of a secret order.

To keep a mystery active requires not only suspending the revelation of the thing itself but investing the gestures of concealment with a mysterious weight; for a mystery is as much its outward dissemblance as some inner essence. A mystery that moves us in this sense dissembles a surreal outer world in order to safeguard the true reality within, and the secret-breaker does as much by making fun of the outward aspect—the rites of dissemblance—as he would in whispering the truth of an aphorism hidden behind the veils.

Imagine, for example, the mysterious essence of humanity hidden away inside of a babushka doll: Pull apart one shell of the painted, wooden woman, and look—there is another of the same form but smaller and painted slightly different but in any case not the singular exemplar of humanity as such. We know this because this one is not exactly like the last, and yet both are alike enough to indicate some common origin—perhaps inside. . . . There is some satisfaction to the process of unpacking human types to their smallest origin, which appears finally like a child inside of the mother: a solid center of being that supports the shells of successive stages of development. But by then of course the game is over, and one is left with little more than to put the toy back together. But what if the core weren't so solid; what if it were something more malleable and altogether fragile, say like the baby doll's form in chocolate, so that when the center is found it could be eaten—then one would have a real mystery left behind in its absence. And the interrogators could ask: Who ate the babushka baby? And the philosophers could wonder: What essence do we no longer have left? Such a transgressive act of auto-cannibalism would be one way then to step outside of the bounds of an otherwise circular game, with the *and this and then this and then this. . .* and the *let's do it again next year* of ritual formations reduced to outward, empty formalisms—with the supposed center gone absent.

The great paradox here is that the mystery remains in force only through its own transgression. Alcibiades must be executed for his profanity—the abyss needs a few living stones thrown into it in order to make a sacred precipice of a pit in the ground, to confirm the open value of its absencing of an absence. If, for instance, it is a coin that is thrown down into the well, then its proscribed value too must be emptied out, so as to be, as in the custom, replaced with a wish. Certainly the value of a wish is not one to which a denomination could be easily attached, neither in pocket change nor otherwise. So the nominalist value of the coin is nullified by its exchange for a wish, which it then becomes the medium for. Nullity thus stands in as equivalent to substance. In the case of the sacrificial human the negation is automatic and absolute; the crime may be indexed, but since a person cannot be, the discrete measurement and recording of the human presence become possible only after the person has been made into a public absence.[37] To be exiled from the human community is thus to be entirely absorbed by it, so that by going outside of its own humanity, humanity defines what it is at the margin what it is not. How to hide this secret source of all identities? And how will a society hide its constitutive lie from itself? The perennial strategy is: Tell the lie in public to everyone but oneself. Thus, "The disclosure of the commodity's secret was the key that revealed capital's enchanted realm to our thought—a secret that capital always tried to hide by exposing it in full view."[38]

If it is true that the world is as we think, then what ought to be acknowledged is that our patterns of thought are not as self-evident as they may appear: that those conceptual patterns likewise constitute the realm of appearances, and that there are constitutive gaps that open up within operations of imaginative reason as it goes about cobbling together this proposition and that. The real mystery is how we come to habitually understand the world as a more or less smooth flow of appearances. Forget Hume's dying fire; how is it that we are not shocked by the sight of the sun again each morning? If causality is a habitual formation of the mind linking events of consciousness into series, like a fast-twitch, channel-surfing eye, with the concept of succession describing the most rudimentary principle of reason, then from whence the tendency to recoil at Heraclitus's aphorism (which I think is intended at least partly for shock value) that "The sun is new again every day"?[39] If the world is punctuated into successive moments by each of our blinks, then the world's turning from sun-lit side to darkened night sky can be thought of as a kind of a planetary blink. Go ahead then: Change the channel on the sun. Close your eyes and see it again in a different light—as fire, as fusion, as circle, as god. If the attributed form is intelligible in itself as substantial only inasmuch as it arises out of sustaining field of flux, then one can say little of reason in itself except that it is one habit of the mind among many.

Even when reason is valorized as the most important cog in the mechanics of the human mind, it can be justified only in hypothetical terms, as in *Let us act **as if** reason distinguished human being*. A sense of nobility may remain, but there can be no reason for the reason itself, precisely because it is viewed as a method—as a way of thinking. In its noblest workings, as in Kant, reason may be employed to justify the ethical treatment of human beings as "ends in themselves," but it remains a hypothetical reason, since there can be no reciprocal reason why one should be ethical in the first place. By comparison, the circular logic of self-serving reason tends to seem that much more substantial—at least it can give an account of itself, even if it is so redundant an account as *I want what I want*, whereas the valorization of reason over desire requires a jumping off from this closed circle, if only to retain the outward shimmer of transcendental thinking. So while one may know that this is what it is *like* to be ethical—to treat others as one would treat oneself—it remains a likeness of a likeness of a likeness . . . , with the moment of transcendence reduced to the willful self-forgetting, so that one only knows how others ought to be treated when one consciously detaches from how one would like to be treated. Short of a presumed reciprocity between one's actions toward others and their reciprocal reactions, and in the absence of correlation between the shape of reason and the world considered as a whole, one is left with a momentary disconnect in the operations of the mind, a jump over that microscopic, synaptic gap inside of every thought by which the person is imagined into wholeness and the world taken in. Within this interiorized conception of the otherness of others and of nature, reality is indexed by its surreality, with familiar phenomena washing over the well-adjusted mind, while the wondrous and the awesome hits us in the face as substantial just because our conceptual apparatuses can find no categorical place for it.

This would be a way through the subjectivism that has made issues of identity with others such a difficult business within modernity. In Rousseau as in Hobbes, things themselves—the world out there, or nature prior to the self-divisions of subjective perception, or communal tribes pre-existing feverish self-interestedness—are phenomena that can only "seem" to be in themselves. And because they belong to the realm of seeming (ultimately pertaining to the *verum factum* that the self makes for itself) forming a picture of these others is a matter for art, rather than reason to accommodate. In the absence of a structured and differentiated whole to begin with (that is, outside of the pre-logical formations of myth[40]) poetry then stands in for reason as a mode of telling one element or drive from another. So likewise, if we cannot know nature as such then we can artistically imagine it. A similar insistence of poetic intuition versus rationality pertains to the barbarian, that most figurative of categories.

MYTH IN MODERNITY

While it has become customary understanding to understand barbarism in terms of strict Otherness, as antithesis to one's own civilizational thesis, this thesis/antithesis logic of identity construction is actually, I would suggest, itself culturally specific to contemporary civilizations that have common origins in the modern European Enlightenment.

Within the context of the modern enlightenment project, civilization is taken to mean the systematic application of a techno-scientific reason, toward the realization of efficiency in relations between human beings and nature, while bringing a measure of transparency to human relations. Within this conceptual/cultural context, which can be traced to such historicist thinkers as G. W. F. Hegel and Immanuel Kant, a barbarian is irrational in the sense of lacking an effective reason (they may have their other reasons, but these fall outside of the trials of testability and managerial transparency). The European Enlightenment man's barbarian is imagined as the opposite of his own clarity and perspicuity in thought: the barbarian who is emotional rather than calculating; the creature of dark amorphous drives rather than clear and distinct reason; and historically speaking, as a backward element opposed to the progressivist drive of modernization.

Toward a foreground account of world history leading to universal enlightenment, Hegel describes the ways or methods by which human beings make the world their own in thought, that is their modes of historical consciousness. In what is perhaps for us (post-moderns) an uncomfortable emphasis, Hegel takes pains at the outset to distinguish the progressive unfolding of historical consciousness from the mythic mindsets that preceded it—a relationship that, by his telling, must appear rather incidental. Progressive history thus begins with a radical disjuncture from the integrated world of myth, that is with the alienation of subjectivity from the world conceived as immediate presence. If Hegel's history ends with the reconciliation of subject and object, or form and substance, or human being and nature, it cannot begin until those separations are first made:

> The historian binds together the fleeting rush of events and deposits it for immortality in the temple of Mnemosyne. Myths, folk songs, traditions are not part of original history; they are still obscure modes and peculiar to obscure peoples. Here we deal with peoples who knew who they were and what they wanted. Observed and observable reality is a more solid foundation for history than the transience of myths and epics. Once a people has reached firm individuality, such forms cease to be its historical essence.[41]

For Hegel, there can be no historical consciousness in the absence of a consciousness of the perspective of the story-teller, with the myths having been passed on from generation to generation and owing their credibility to the others who have told them before. For where does the myth-maker stand within the narrative? How can he tell the stories he tells? Homer must begin his myths with invocations: "Sing Muse!"[42]; or with an entreaty: "call to my mind, gods."[43] The myth-maker gives up his voice as a subject with a unique perspective so that he can become a medium for the gods, with there being no strict separation between self as speaking subject and a world of objects. This compact unity of myth mirrors the political unity of the tribe, with there being no strict separation between individual and community, with the community joined by common ancestry and a genetic-bond. Myth for Hegel is pre-historical in the sense that history hasn't gotten started yet through the separation between man as subject and a world of objects, with history beginning with self-consciousness, and driven toward the full realization of self-consciousness in freedom. This lack of rational self-consciousness is the basis of barbarism for Hegel, casting peoples as being somehow outside of the stream of historical progress because they do not see themselves as being set apart from the world that they give voice to in myth.

Myth is summarily rejected by Hegel as an expression of full consciousness because there does not appear to be a human subject at its origin. Most significantly, myths do not come to self-presentation as the works of human beings, and it is easy to see why not, especially when they include cosmogonies or narratives that include secret negotiations among the gods or bird's eye views of human affairs. Homer for one (if he is indeed just one) is explicit in self-denial of his capacities as an author, invoking the Muses,

> —for you are present,
> And know all things, while we
> Hear only reports and know nothing[44]

And yet isn't there a glimmering sense of an emergent self-awareness in this epistemological impasse, later to be radicalized in Socrates's dictum (in turn inspired by a failed attempt at refuting the Oracle at Delphi) to know that one does not know, and given experimental clarity in modern science in the denial of any real knowledge of causality or the essence of beings as such? What Hegel would deny of such an evocation of untold subjectivity is that Homer's impasse points to an impossible gap between perceptual and cognitive capacities and the sources of knowledge, and hence requires an external referent that is thought to reveal itself in some immediate way to the storyteller. Homer takes this most distant of perspectives—the view that the gods on Olympus have—and internalizes it as a voice of inspiration. However much we, as readers or listeners may be made to feel present in the action of the epic, like a half-way divine creature ourselves, experiencing the pure presence of enthusiasm in hearing the story, by the end of the myth we know nothing about the author, who vanishes into the tale retold.[45]

Whereas the first person accounts of what Hegel calls original historians—the first of three unfolding modes of history in Hegel's system of progressive sciences—the method of first-person recollection—seem to possess an immediacy in presentation, it is still a representation in that it necessarily involves the perspective of the historian. So Thucydides, for example, announces his presence in his *History*. . . in the first person "I." Properly speaking, historical consciousness must begin and end in a self-conscious that is both subjective and objective: that is, the self as object of its own subjectivity. In the first instance this is represented as a subject mediating the objectivity of his personal sense experience, wherein the self is revealed in a limited way in the objective determinations of perspective—the bias of the author—a proviso not necessarily admitted by the authors of myth, who conventionally take on a collective identity as they become mediums for stories that no one person could tell, taking in an epic narrative arc that no person save a god could see all in one view.

Because of this paradigmatic conflation of subjectivity with both the world of nature and the sacred world, dialectical thinking can get no toehold on myths—they are simply too cyclical, with every unfolding of successive moments a simultaneous retreat of the same. The same thing stands out and is hidden; every dictum is an aphorism; every separation implies some deeper identity that the openings of distinction fall back into. So, for example, Athena, having sprung from the temple of Zeus, is her father's favorite daughter[46]: She is the ratiocination of the patriarch—his plan or yet-unspoken design—"*muthos*"—for the future—sharpened into a feminine instrument, as if Zeus were to become an intelligent, outspoken young woman. And as the ancient story implies, he must have had it in him. The fact that the father and daughter fight with each other, far from disjointing the identity between the two, gives voice to a signal pattern of externalized selves, wherein we confront otherwise unmanifest modes of our own subjectivity.

But aren't we going back too far here in ascribing an immanent eruption of subjectivity, conditioned by its own externalized lack, back onto myth itself, that closed circle of naturally regulated desires, tribal continuities, and epic, whole-cosmos catalogues wherein everyone knows their place? This is the way that myth is often portrayed, as a naturalized and long-lost model for integration. One can imagine such a relation linking Hegelian progressivism and Greek myth: that somewhere in-between subjectivity intervenes, the organic *polis* and natural philosophy of the golden age of Greece are upset, and Western civilization begins its long voyage past the artificial, horizonal limits of the River Oceanos and out into the horizon of the whole Earth proper, to return again, though on a more solid basis of subjectivity that knows itself, rather than borrowing its sense of substance from stories inherited from myth and ultimately passed off as the works of ancestral gods.

And yet, if we read the mythic pre-history of the Greeks with a sensitivity to the constitutive relations that held their *kosmos* together, it is fairly evident that there are gaps even there that open up between selves and selves—say wherever individuals understand themselves as they are seen by others (as is very common) and yet perhaps not always as they would want to be seen. There is room for unfulfilled desire even in a world so rich in interrelations such as we find in the Achean beachhead camp, or Odysseus's castle, or the assembly of gods on Olympus. All is not perfect in myth; contra Hegel, far from having their individuality suppressed by a seamless weave of traditional formations, the actors in myths must constantly struggle to assert their sense of selfhood by projecting it onto the perceptions of others. If in our reductionist, biologically driven times we may say that we are what we eat, for Homer's heroes we are what we are seen eating. So where a hero sits at the table, what cuts of meat he is offered and what he does to deserve those distinctions constitute his being. If the Hegelian concept of subjectivity is constituted by an absence at the center of the self—an undetermined, hence free being that can fill in its purposes as it likes—by comparison the subject in myth comes into public presence in their relations with others, with the opening of the self projected out into the space between the self and others. And as anyone who has spent some time with others knows, these projections and reciprocal projections admit a variety and flexibility that one person would have trouble dreaming up by themselves. If we can say that there is a subjectivity in myth, it is one that is not so easily isolated within the cranial cage of one's mind but depends rather upon public places of disclosure and hence the presence of others.

BIG OTHERS AND OTHER OTHERS

Whereas the big "Other" is quarantined from the rationalized modern order, the Other has a tendency to be valorized by post-modern critics who intuit certain gaps in the modern self-understanding, with the imagined, anti-enlightenment barbarian, who says *rational thought is just one habit among others* prefiguring a critique of modernity arising from out of the assumptions of modernity. Stepping back to the initial ethical proposition that a civilization's treatment of its barbarians can be taken as a clue, if not to these others themselves, then to the civilization's own substance and qualities, what an ethics of difference and linked ontology of self and other can be taken to imply is that our conceptions of others register our own sense of self-respect.

APPLIED BARBARISM

Barbarism apparently still has its uses, though even when it is taken in hand, as one would hold a hammer in order to strike something with it, the ideology of barbarism maintains its quality of working un-self-conscious determinations onto whoever employs it. Or in other words, barbarism is not a neutral rhetorical tool that can be used in a rude and uncomplicated way to degrade others but as a lens of perception the ideology has its reciprocal effects. So although barbarism may be used rhetorically to condemn another's actions, for example, poor table manners, like eating with one's left hand, or egregious acts of violence, like the targeting of civilians in military campaigns, still, whenever someone says *You're barbaric for doing that . . .* there is always a certain naive ignorance implied. In this sense to deploy the language of barbarism is to reveal a certain lack of self-consciousness, such that the undetermined freedom that exists inside of every rational, ethical agent is filled in with unconsidered prejudices, automatic fears, and a blind, instinctual self-assertion that nonetheless cannot arrive at self-knowledge precisely because the core of the self has been objectivized and externalized in the form of a mortal enemy. This is a picture of a self at odds with itself, fighting with its own shadow in a weird compulsion of auto-rejection—a fight that is then only fully reconciled through suicidal, mutual annihilation.

In instances of what "we" would automatically tend to agree are acts of barbarity—like, for example, a story from tribal regions at the fringe of the state laws of Pakistan of the burying of a woman alive for disobedience[47] —what is registered is the evidently unenlightened, crude domination of others (in this particular instance, of woman by man) perhaps for nothing more than the immediate, unreflective inclination toward domination itself and for its own sake. In such instances, the properly reflective, and free response to such egregious acts of violence against humanity—of the victimizer against victimizer as well as victimizer against victim—is to simply wonder why. To ask why was this done? And the more determined, resolved, and stupid the excuse for an answer may be—*she was asking for it* or *they are evil* or *I wanted to*—what saves human freedom even in spite of its apparent abrogations is the freedom to think that things could have been otherwise, that the historical disasters of slavery, of commodification, mass extermination, the redaction of human beings/suspects from public knowledge, etc., only appear as necessary after the trespasses have happened. And even then, one has to wonder.

For instance, what if, even under the conditions of a strict materialist determinism of some sort—social, economic, institutional, psycho-chemical, genetic—there are these momentary unpredictables—like static interrupting regularly scheduled programming—and that in their turn these interruptions take on an agency of their own, nonetheless determined, but determined quite differently by an underlying cause that is itself unpredictable. So, even if the barbarian acts as an automaton of conditioned desires, a kind of clockwork of appetites and unthinking aversions, half-masked from its own critical capacities with imaginative fundamentalist beliefs, there remains the question of when the automaton was turned on. No machine is a closed system; for no matter how finely tuned the gears or extensively and meticulously programmed its functions may be, there always remains the question of the exogenous environment within which it operates, including under what specific conditions it was activated. Marx suggested that men make history, though not under conditions of their own choosing.[48] It is this materially conditioned status of action that holds out the promise of reconciliation while admitting the contingent character of that possibility.

What does this mean for how we see others and judge their acts as barbaric? The judgment itself is not necessarily blameworthy; even pre-judgment might in some (quick) circumstances be justified. What is at issue is the determination of judgment by those circumstances, to the extent that it loses its qualities *qua* judgment: as in its flexibility, its recognition of its own limits, and its specific conditions as they inform the possibility of being able to make a free judgment in the first place—i.e., in *that* place. All of this is given up, rather circumspectly, as soon as one assigns that identity of barbarian to another. Then one says—*I know why you did this—it's because you are backward* or *evil* or *irrational*, etc., whatever empty signifier may be called upon to ignorantly fix a person or a people in place. And as if to confirm the behavior, with the assignment of the title of "Barbarian," the exiled Other begins to act the part; for what possible response is there to such an assigned identity but a *No, not that, this . . .* thus countering the assigned identity with a reciprocally determined statement of alterior identification? At that point, a properly Socratic answer of "I don't know," would have no discursive grammatical or logical toe-hold in the exchange between accuser and accused. From *You're a barbarian, I don't know* doesn't follow, though the latter can serve as an answer to a "Why did you do this?" (or more authentically, in a fuller spirit of mutual recognition, a "Who are you?). At moments of outrage, it is this genuine questioning that offers, if not reconciliation with the acts of outrage (the event itself is dumb) then at least provisionally the potential to accept the humanity of the most seemingly inhuman others, ourselves included, so as to save ourselves from barbarism contra barbarism.

In simple terms, what this reciprocal pattern of projected identities means is that the real barbarians are those who see others as barbarians. This principle offers a code for understanding aggressive political rhetoric and its terms of derogation—and all too well, apparently. What is striking in this sense is how easily reversible political insults are, with their meaning consisting of a relation between intention and unintended critical self-reflection. The meaning of a political accusation thus lies in between an assumed self-identity and an unrealized projected identity. The object of indictment can readily be mapped back upon the person who makes it, as in for instance, this ignorantly sage observation that,

> The gravest danger our Nation faces lies at the crossroads of radicalism and technology.[49]

. . . a danger observed from a perspective of radical conservatism and technological predominance. If there is a charge of ethical equivalence to be made here, it belongs exclusively to those who call others barbarians. Keeping within the limits of critical self-knowledge, which is never an *a priori* given but rather must unfold discursively, the kind of reflection that accepts responsibility for the world conceived as a web of constitutive relations need not imply universal culpability and some assumed fallen nature of political man. (That may or may not be the case—at any rate it is beyond human powers either to abstract political life to some sort of pre-discursive, prehistorical Edenic bliss and fall from grace, or to take a god-like perspective that would be needed to assign universal guilt.) Rather, what the reciprocal nature of political discourse can be taken to mean is much more accessible to common reason, as in something along the lines of the taunting playground rejoinder: *I'm rubber, you're glue! What you say bounces off me and sticks to you!* If that's the childish response, then what kind of childish gesture originally caused it?

What this means is that understanding does not begin by choosing sides, but is initiated rather by a leap of imaginative reason that puts one into the ethical conscience of another—what must also be a hypothetical ethic in that it always considers the possibility of its own position not being true. For it may be, as Hobbes argues, that human beings are rational merely in the sense of robots—rational automatons, like the machines disguised in hats and cloaks that Descartes sees in the street—conditioned by appetites and aversions, with an ethical public life dismissed simply as impossible, or even dangerous.[50] For what is an-other but a creature of these hard-wired interests, with a will and reason of its own for sure, though distinct as the will of another is distinct from one's own will and set apart inasmuch as their reason is a tool used to serve their interests, not one's own? Why then would one treat others as having value in themselves, and not as automata conditioned by externalized reflexes? What is really human in the human being that is anything more than this *animal rationale* (with reason reduced to self-interested instrumentality) hence primarily animal? At the root of the ideology of barbarism is an inability to make this imaginative leap, understanding others not simply as creatures of instinct and interests (so as to ask—*What will they do next?*) but as intelligent and ethical creatures like ourselves.

This is not to say that all of us are incapable of irrational or unethical behavior. There are many reasons to trespass on or otherwise disrespect others. What though are the reasons for peaceful settlement and identification with others? A helpful first step would be to realize the essentially relational sense of identity formations, even and especially in those instances of violent antimonies.

THE BARBARIAN AS VANISHING SUBJECT

When we think about the end of subjectivity as the determination of human agency and locus of freedom, the Western tradition seems to tend in its culmination toward the emptying out of the subject into a *nihil*, an unmediated abyss at the center of human being, with the reality of human existence confirmed precisely by that absence; since if a being is to be, it must necessarily take on certain determinations this way or that—for example, if a being were to take on a color, and to be, say green, like a fern is green. That fern then is not a free being, inasmuch as its being is essentially bound up with its phenomenal qualities. Hegel distinguishes that kind of objectification of beings—he uses the example of a rock bound to the earth by gravity—from the subjective form of freedom that all human beings ideally should enjoy.

This freedom of the free subject has at its core an undetermined essence; that is, a free being is a being that could be otherwise. In humanity, this notion of freedom is attenuated by its being put into practice in thinking, that is when we are free and conscious of it. For it is entirely possible that one could be free and not know it, or likewise, to be unfree and not know that. When consciousness is added to freedom, freedom necessarily makes its demands known, for no one can truly know that they are free and yet accept their unfreedom as some essential determination of their being.

As human beings, in whatever one's conditions may happen to be, we possess the open potential to know our being free; though once acknowledged, one may still forget, so as to remember again, and again. If freedom is a negation of a negation, as Hegel suggests, then it is a nihil that fills in its own content through creative fits of forgetfulness, undoing its own conditions of possibility so that its pasts may invade its future in unconditioned ways. This is freedom as originality, where to be free is to be original, in the sense of what is unprecedented. This is what freedom means as a progressive ideal, where the capacity to act freely into the future begins with a break in the lockstep logic of historical causality, where every action is really a reaction, conditioned by interests, upbringing, and pre-existing circumstances.

A free gesture, waving away these derogations of the free spirit as incidental to human being as such, could be initiated only by an historical being that is nonetheless outside of history in some way. In past traditions this was treated as the transcendental, eternal *psyche*, as in Plato, or as individuated soul-substance, equal before a monotheistic God, as in St. Augustine's Christianity. In modernity the locus for self-consciousness and free being became the subject, a worldly being, capable of exerting power, and in some ways describable as the center and basis of power and the origin from which every (political) push—or effort of valuation—ultimately occurs. In early modernity (as in Hobbes) this was the individual subject recognized as citizen before the authoritarian state; in late modernity (Hegel, Kant) this was the

subject as equal before the law. In the aftermath of modernity it is this billiard ball model of subjectivity that breaks down into a flow of identities in transit and fractured signifiers. As the subject comes apart, leaving off the lawful and communal aspects of Hegel's free subject at the end of history—the last man who enjoys his laws and his state as his own is left to become the subject considered as a negation of a negation—what is the absent self.

The undetermined center of a human being is what a schoolman would call its quiddity or its whatness: The real being that predicates can then be attached to, like tails pinned onto a donkey. The freedom of a human being then is just what cannot be pinned down with a predicate; it is what they would make of themselves, or what is in conscious, practical terms an appeal against its own objectification.

And yet, the recognition of others remains persistently challenging, even after the progressive ideal is offered up of a subject liberated of objectifications and assigned instrumentalities, since what remains is as black and colorless as the pupil of another person. For even in those moments of humane rapture in identification with others, when one looks another person in the eye and trusts to have known their being just by their looking back, at the origin of that gesture, behind the reciprocal gaze there is the free thought that is capable of being changed.

The fragility of this condition of free thought is readily apparent. The freedom that one can see in the eye of another human being gets determined even simply by its being focused on one particular thing. As soon as sight singles out another being, it is bound, in a sense, to that being. The gaze, so it is said, is "fixed" upon a particular thing that has become its object of care. In that attention, one begins to belong, reciprocally, to that being at the center of one's attention, as a moon may seem to belong to its planet. In this way the gravity inherent in worldly being takes its hold on what should in essence be a free being. Because of this apparent insistence of the necessity of belonging, by which for example nations or races of people may gather in discriminatory enclaves, freedom finds its limit in the existence of "us" and "others." It was in such a spirit that Hegel wrote,

> One cannot skip over the spirit of his people any more than one can skip over the earth. The earth is the center of gravity; a body imagined as leaving this center can only be imagined as exploding into air. So it is with the individual.[51]

To see beyond those apparent limits and the fixity of our worldly being, humane attention may be fixed only upon what is itself unfixable, in precisely the kind of gaze of mutual recognition that is the usual basis of civilized, friendly behavior with others, where, in a simple look of mutual respect, the undetermined, free self projects itself onto another undetermined self and sees its own freedom—which should by essence be unrepresentable—in the other's eyes. In that gesture of respect the "they" become as unrepresentable as the "we" or the "I" would hope to be.

There are many essential freedoms—to movement, to habitation, to speak or to be silent, of livelihood and cooperation—but essential to all of them is the free capacity to change one's mind. Freedom to associate—that is freedom to choose to cooperate with whomever one chooses. Freedom to livelihood—including the freedom to be poor and live well. Free habitat—to change homes and be assured of a place to live. Free movement—universal mobility rights, the freedom to change one's mind and simply move. Free speech—the discursive freedom of engaging in dialogue, ready to be convinced by others. Freedom to silence—the right not to tell or to answer, where the right remains intact even after it has been given up before. At the center of all of these essential rights is the free-thinking subject, who by their very essence must demand as much. That is, to be conscious of one's own freedom is both to see it in others and to see the freedom of self and other as necessary. For a solitary free self cannot open his eyes and still be free, unless there is another free self in whose eyes one can observe the limitless depth of another undetermined being. Freedom has its insistence in the perspective from which any constraints that intervene must take on the character of incidental, non-essential situations. What makes this progressive, reality-altering perspective possible is the de-centered self, uniquely capable of seeing its own essential absence outside of itself in another.

NOTES

1. The Confucian scholar Yong Huang proposes the ethic in his essay "The Ethics of Difference in the *Zhuangzi*," *Journal of the American Academy of Religion, (2010) 78* (1): 65-99. Huang proposes a cross-cultural ethic along the lines of the golden rule with a reflexive twist—that one treat others as *they* would like to be treated—a dictum that raises the baseline observation that ethics begins with difference, for if there were some sort of universal to begin with, there would be no need for an ethos or a way in between. Hence the search for that alternate original in the other who was here first. We begin with our own differences to be confronted by the differences of others, wherein what may at first seem like a singular identity comes to appear as a manifold, a sharing of what one lacks in oneself with the other.

2. The liberal tradition takes this separation of peoples to its radical conclusion in Rousseau, who idealizes a people:

> . . . that is already bound together by some tie of origin, interest, or agreement . . . ; one that has no deeply rooted customs or superstitions; one that has no fear of being overwhelmed by a sudden invasion, and while remaining aloof from the quarrels of its neighbours, is capable of withstanding any of them alone, or of enlisting the aid of one to repel another; one in which each man can be known to all, and in which there is no need to place a larger burden on anyone than he can bear; one that can do without other peoples, and that all other peoples can do without.

Jean- Jacques Rousseau, "The People," *Social Contract*, Book 1, Chapter X, *The Essential Rousseau*, Lowell Blair translation (New York: Meridian, 1983), 44.

3. Jürgen Habermas, *The Divided West*, (Cambridge: Polity, 2006), 12.
4. *The Divided West*, 11 (The emphasis is his.)
5. William Butler Yeats, "The Second Coming" www.potw.org/archive/potw351.html.
6. Slavoj Zizek, *On Belief*, (New York: Routledge, 2001), 1.
7. *On Belief*, 84.
8. Hannah Arendt, *Between Past and Future: Six Exercises in Political Thought*, (New York: Viking Press, 1963), 170.
9. "This is the meaning of the famous dialectics of freedom and necessity in which both eventually coincide—perhaps the most terrible and, humanly speaking, least bearable paradox in the whole body of modern thought. And yet, Hegel, who once had seen in the year 1789 the moment when the earth and the heavens had become reconciled, might still have thought in terms of the original metaphorical content of the word 'revolution,' as though in the course of the French Revolution the lawfully irresistible movement of the heavenly bodies had descended upon the earth and the affairs of men, bestowing upon them a 'necessity' and regularity which had seemed beyond the 'melancholy haphazardness' (Kant), the sad 'mixture of violence and meaninglessness' (Goethe) which up to then had seemed to be the outstanding quality of history and of the course of the world. Hence, the paradox that freedom is the fruit of necessity, in Hegel's own understanding, was hardly more paradoxical than the reconciliation of heaven and earth." Hannah Arendt, *On Revolution*, (London: Penguin, 1990), 54.
10. *Politics*, 1280a.
11. *Iliad*, 12. 320-332.
12. *Republic*, 390a-391c.
13. Consider Mircea Eliade's earnest efforts to understand myth in its own terms through the interpretation of its symbols, in *Cosmos and History: The Myth of the Eternal Return*, William Trask translation (New York: Harper, 1954), and *Myth and Reality*, William Trask translation (New York: Harper, 1963)—along with Ernst Cassirer's neo-Kantian approach to considering myths as categorical deep structures of the mind in *Language and Myth*, Susan K. Langer translation (New York: Dover, 1946), and *The Philosophy of Symbolic Forms, Vol. 2: Mythical Thought*, (New Haven: Yale University Press, 1996).
14. "The *Imitatio Christi* proper to the dual knowledge of *Homo Dei* must be to recognize the personality of the god or goddess Eros-Amor, Kosmogonos, not where it can be neither sought nor found, 'out there' somewhere, in transcendence, but—as Christ did—in oneself. And not oneself alone, but all things, all events: in every individual, just as he is—crude, fine, or superfine—God's mask." Joseph Campbell, *The Masks of God: Creative Mythology*, (New York, Penguin: 1968), 647.
15. Thucydides, *History of the Peloponnesian War*, Richard Crawley translation (Vermont: Everyman, 1993), 1.6.
16. The meaning of an outwardly imperial attitude toward another consists in the enlightened attitude even more so than the violence that promulgates it. What results from the primary epistemological violence is an imperium that projects a singular identity onto others, with the barbarian rendered speechless by the presumption that the common ground of any pre-civilizational interaction is force. Thus Edward Said remarks that,

> What is striking in these discourses are the rhetorical figures one keeps encountering in their descriptions of "the mysterious East," as well as the stereotypes about "the African [or Indian or Irish or Jamaican or Chinese] mind," the notions about bringing civilization to primitive or barbaric peoples, the disturbingly familiar ideas about flogging or death or extended punishment being required when "they" misbehaved or became rebellious, because "they" mainly understood force or violence best; "they were not like "us," and for that reason deserved to be ruled.

Edward Said, *Culture and Imperialism*, (New York: Vintage, 1993), xi.

17. Joel Kovel, *The Enemy of Nature: The End of Capitalism or the End of the World*, (Zed: London, 2007), 95.

18. Luce Irigaray, *Speculum of the Other Woman*, translated by Gillian C. Gill, (Ithica: Cornell University Press, 1985), 213.
19. Thomas Hobbes, *Leviathan*, 85
20. Hannah Arendt, *The Human Condition*, (Chicago: University of Chicago Press, 1958), 10.
21. Jean Jacques Rousseau, "Discourse on Inequality Among Men," Preface.
22. "Discourse on Inequality Among Men," 176.
23. Rousseau asks the attentive reader to understand "why our needs and pleasures take on new objects; and why, original man having vanished by degrees, society no longer offers to the eyes of the sage anything but an aggregation of artificial men and factitious passions that are the work of all these new relations and have no real foundation in nature." "Discourse on Inequality Among Men," 199.
24. Thomas Hobbes, *Leviathan,* (London: Penguin, 1985) 139, 182-183.
25. "Let us not put ourselves in a fictitious primordial state like a political economist trying to clarify things. Such a primordial state clarifies nothing. It merely pushes the issue into a grey misty distance. It acknowledges as a fact or event what it should deduce, namely, the necessary relation between two things for example, between division of labour and exchange. In such a manner theology explains the origin of evil by the fall of man. That is, it asserts as a fact in the form of history what it should explain." Karl Marx, *Economic and Philosophical Manuscripts of 1844*, in *Writings of the Young Marx on Philosophy and Society*, Loyd Easton translation (Garden City: Double Day, 1967), 289.
26. "Today humanity seems ready in a new way to assume its own limitedness and, despite the insurmountable particularity of the knowledge that science comes to know, to find satisfaction in the increasing mastery of nature that it owes to that particularity. It even takes into account the fact that with the increasing mastery of nature, the domination of human beings over human beings is not eliminated but, counter to all expectation, becomes ever greater and threatens freedom from within." Hans Georg Gadamer, *Reason in the Age of Science*, Frederick Lawrence translation, (Cambridge: MIT Press, 2001), 3.
27. Friedrich Nietzsche, *Beyond Good and Evil*, Kaufman translation, §67
28. Plato, *Symposium*, 177d.
29. *Symposium*, 215b.
30. Jean-Jacques Barthelemi, *The Travels of Anacharsis the Younger in Greece During the Middle of the Fourth Century before the Christian Aera*, 2nd edition translated by Duc de Nirvernois (London: Vernor and Hood and Lackington, Allen, and Co., 1798) Vol. 2, 398-9.
31. *Travels of Anacharsis*, Vol. 2, 391. This profusion of inscriptions gives evidence of the Greeks' profound sense of public space, decorated with monuments to the spiritual imagination and cues for collective reflection.
32. Friedrich Nietzsche, *The Gay Science*, Walter Kaufmann translation (New York: Vintage, 1974), 1.16.
33. Hannah Arendt, "*Zur Person*" interview. http://www.youtube.com/watch?v=dsoImQfVsO4@12.45
34. *Travels of Anacharsis*, Vol. 5, 213.
35. *Travels of Anacharsis*, Vol. 2, 463.
36. Sophocles, *Oedipus Tyrannus,* Francis Storr translation. www.perseus.tufts.edu/hopper/text?doc=Perseus%3Atext%3A1999.01.0192%3Acard%3D1329
37. This is the singular thematic of Arendt's description of totalitarianism in its radical fulfillment, wherein the "subterranean stream of Western history has finally come to the surface and usurped the dignity of our tradition," with the result being the redaction of people into nothings. [*Origins of Totalitarianism*, xxxi] From that tipping point of modernity on, "Only with a completely organized humanity could the loss of home and political status become identical with the expulsion from humanity altogether." [*Origins* . . ., 297].
38. Giorgio Agamben, *Means Without End: Notes on Politics*, Binetti and Casarino (trans.) (Minneapolis: University of Minnesota Press, 2000), 75.
39. Heraclitus, fragment 32.
40. The connection between the process logics of reason and elemental, mythic thought formations is drawn out first in Vico, then again by Eliade and Cassirer, all of whom relate the

substance of myth and certain deep resonances of the human historical spirit to the outward forms of reason.

41. G.W.F. Hegel, *Reason in History,* I.i.

42. Homer, *Iliad,* I.1.

43. *Iliad*, 2.521.

44. *Iliad*, 2.523-25.

45. It is noteworthy that the most stylized elements of Homer's poetry—his invocations, similes and repetitive epithets—that may seem clunky when the poems are read silently sound precisely right when they are performed aloud in public and from memory. One translator, Stanley Lombardo, whose versions of the epics were formulated from mnemonic notes for public performances calls attention to the sharpening of attention that these ritual formations introduced in his retellings of the poems:

> In performance, I found myself isolating the similes somewhat and marking them—pausing a little before and after, changing the voice, dropping any percussion I may have been using—in order to bring out their quality as poetic events distinct from the poetry of the narrative and the speeches. I found that the narrative resumed with a kind of quiet power after a simile had been given full attention in this way, and that the audience's engagement with the performance was deepened. [*Iliad*, Translator's Preface, x-xi]

It is as if a full involvement of the audience in the story requires some reminder of the perspective from which it is seen, a coming up for breath before the next episode. This need to punctuate the flow of phenomena understood as narrative is thus filled in, if not by a self-presentation of author himself, then by a self-conscious display of style, pulling focus from inner flow of the story to its outward forms of presentation, and opening up a place for the subject—both as teller and as listener—as a medium in between.

46. *Iliad*, 5.785

47. http://www.ahrchk.net/ahrc-in-news/mainfile.php/2008ahrcinnews/2207/

48. Karl Marx, *The Eighteenth Brumaire of Louis Bonaparte,* in *Karl Marx and Frederick Engels: Selected Works,* (Moscow: Progress Publishers, 1968), 97.

49. 2002 National Security Strategy of the United States of America

50. Thomas Hobbes, *Leviathan*, 118 ; René Descartes, "Meditations on First Philosophy", in *Descartes: Philosophical Writings*, (London: Nelson, 1969), 73.

51. G.W.F. Hegel, *Reason in History*, 37-8.

Chapter Two

Beyond Barbarism, Love

ALTERITY OR NOTHING!

FORMLESS

> A dictionary would begin from the point at which it no longer rendered the meanings of words but rather their tasks. Thus formless is not only an adjective with a given meaning but a term which declassifies, generally requiring that each thing take on a form. That which it designates has no claim in any sense, and is always trampled upon like a spider or an earthworm. Indeed, for academics to be happy, the universe would have to take on form. The whole of philosophy has no other goal: to provide a frock coat for what is, a mathematical frock coat. To declare, on the contrary, that the universe is not like anything, and is simply formless, is tantamount to saying that the universe is something like a spider or spittle. [1]

What would the world be like ontologically if alterity and difference were embedded in the being of every being? In Bataille's endorsement of the declassifying effects of the formless, the effect of staking a claim on alterity is to call into question the categories by which the world is made intelligible. If it is somehow more realistic to address the being of beings as characteristically indistinct and in flux, as in a kind of primordial soup of becoming, then the cost of this claim seems to be the terms of reference for sorting the world into packets of information that can be communicated to others, precisely the sorts of dictionary definitions that Bataille begins by deconstructing, reconfigured from claims to the thing itself into tasks of ordering by fiat.

This ontological impasse between form and flux has ancient roots. In the Platonic dialogue, the *Theaetetus*, an impassioned student of mathematics and, as it turns out, fluid student of philosophy, Theaetetus is caught up in the dialogue between his teacher of mathematics, Theodorus, and Socrates.

Theodorus presents an indignant, embattled rationality, a sentiment he aims at the followers of Heraclitus—"men of flux," who thought that underlying reality was like a river, constantly changing, constantly flowing from one set of appearances, one set of tensions, to another.[2] Because these Heraclitean motion men see reality as fundamentally about change, consisting in a field of interrelated appearances, of contestation and of violent motion, they see society that way too. So Heraclitus says:

> War is the father of all things and king; some it shows as gods and some as humans, some it makes slaves and others free men.[3]

This kind of radical relativism, one that would melt the distinction even between the divine and the human in claiming that all categories are the products of violence, goes back even further than Heraclitus and what Socrates calls the modern thinkers of his time to Homer, and a society underwritten by explicit endorsements of violence as the basis of shared reality, as in Odysseus's discipline of the critic Thersites, or that Zeus's will be done in the end.

This would in many ways be an automatic society, with the critique going so far that it loses its critical capacity, coming to a conclusion that all critiques and all truth claims are, in essence, the masks of power (its own critique included). And so, by rights of logic, the really critical thinker should hold off on theories altogether and just give positions; and yet, these motion men would keep speaking. This is what Theodorus doesn't understand, that the students of these teachings whom he saw springing up like mushrooms would carry on from one aphorism to the next, speaking a logic that would undo logics and expressions of power, following their king, Heraclitus, a man who had himself preemptively abdicated from power.[4]

Socrates offers that maybe these aphorisms were esoteric teachings, that although perhaps a truth may be contained in a Heraclitean maxim, as in "Just as the river where I step is not the same, and is: so I am as I am not," it is a truth made to shape a perspective, more so than to give that perspective content.[5] That substance would come instead in the form of secret teachings. In this softened, self-reflective criticism by Socrates, the interval between the form of an ontology and the content it expresses and gives rise to would allow for a truth in Heraclitean ontology without going crazy for that potentially caustic truth, which is that things change, and in so changing become other than they are.

In the dialectic between chaos and order, or in their epistemological corollaries of ignorance and knowledge, the claim to universal relativism, like the claim to universal unity (as in Parmenides's theory of oneness) is a partial perspective claiming wholeness. It is a paradox that relativism wouldn't apply its own critique to its truth claim; but as Socrates suggests, every "theory," could be criticized for its partiality, each being after all not an independent system of ideas but a perspective offered by one person, however much a particular view may be parroted by many. And yet, the argument from the limits of knowledge does not itself retreat into relativism; Socrates's argument makes no claim to its being a whole view. The dialogue ends with the enigmatically consistent assurance that the interlocutors know that they do not know, after sorting out the things that surely cannot be known along the way, and from what perspectives certain things certainly cannot be known. Thus Socrates introduces a kind of knowledge described precisely by its limits, in the qualification, "By thinking that the things which he knows are sometimes things which he knows and sometimes things which he does not know."[6]

There is a kind of common sense in the outrage of Theodorus, the teacher of mathematics, toward an ontology that makes a truth of change, relating reality as a chaotic field of happenstance relations. The emotional response from the mathematician and the reasons it gives rise to offer an opportunity to carefully consider his injunction to turn our thoughts away from any inquiry into the formless itself, so that it should not impress its formlessness onto our thoughts.[7] For what kind of thinkable form could it have, short of that which it is other to? To the extent that formlessness or strict otherness is taken up as the object of inquiry in and of itself, there is a certain danger that categories of understanding will dissolve, with identity formations veering toward the eradication of the other, seen as incompatible with what is otherwise arbitrarily identified as essential to a people's collective being.

The value in self-other distinctions consists rather not in the other as such—which is as formless a category as one can imagine, more so even than the test case of mud offered as a challenge to Plato's theory of forms in the *Parmenides,* or Bataille's irreverent conception of the universe as a drop of spit—but rather in how the other is imagined in essential relations with the self or within a collective identity. In terms of projected identities, where we see ourselves in light of our own others, certain intuited but unacknowledged gaps within a self-understanding are bracketed and then transposed onto another who would serve as a foil for one's own fractured self-understanding. So enemies—big Others—are singled out because they embody everything unfulfilled in the antagonizing subject. So if we are licentious, our enemies are hard-nosed ascetics living in caves. If we are an artificial people, then our barbarians become the natural ones, so attuned to reality that they seem subhuman in contrast to the civilized person's objectively divided, hyper-secured being. Others are imagined thus as an identity's non-realized being.

As a mode of reconciliation within the distinction between self and other, though the initial ethical response may be to hold off on these kind of self-other distinctions (and there are many others left to fill in), I wonder if there isn't an opportunity for critical self-knowledge in investigating the precise terms of collective prejudice, toward a more inclusive identity where the self sees itself in its self-made others?

So what if, some good conscience may ask, "we" or "I" have no barbarians? Well then there is neither a "we" nor an "I"; or else, the identity is proposed as universal, since it consciously includes its own opposite in the form of a reconciled, self-less self. The gesture of recognition that signals self-respect begins by dropping the determinations of self, or what is the same, by internalizing one's own alterity in an acknowledgment of the constitutive gaps in self-knowledge.

Yet such self-less selves bursting with the whole, non-identity of being are rare, if only because in them being finds ease with its own non-being. Perhaps we can find an image of this kind of enlightened self-abnegation in the person of Socrates, with his calm composure toward death and grace in admitting nearly universal ignorance. Even Christ had his moment of feeling forsaken, of aloneness in a world alien to his death and desires. Yet for the vast mass of human life the absence that is the substance of the free self remains foreign. In those leftover longings for an unrealized otherness there remains only an echo of alterity, an emptiness from the inside made to sound far away, so that, in screaming at the world, some consolation is found in a world that screams back; all that is other thus serve the inherently partial yet still significant function of sounding back echoes of a fractured consciousness.

In lieu of a recognition of others predicated upon an acknowledgement that the self is what it lacks—and that this is considered universal, in due measure to self-consciousness—the default mode of cultural identification seems to take the formless form of a projected alterity: such that "they" are the not to what "we" are. In this form of identification, the *nihil* at the center of the undetermined self remains, though not in a self-consciously free form, since its open character is excluded. Instead, "they" get it, "they" are the source of "our" identity; though significantly, this means that the other's identity, purely considered, is beyond their own telling, in the sense that they are, in essence, what they were before they were changed by us. "We" become the basis for "their" historical narrative. The point then becomes not merely finding a cultural origin for such a metaphysics of identity in Western historicism (from Christian Eschatology and then German Idealism) but beyond this, establishing recognition that no such identity by exclusion can be a conscious one. Epistemologically, an *I'm rubber, you're glue* form of identity can make no headway; thus "we" cannot get either what "we" positively are or what "they" are negatively. And yet, however much the misunderstanding of both self and other may preside over the relationship, each side turns out to need the other in the worst essential way.

Here what becomes apparent is a self-subsisting collusion by opposition, wherein the undecideable core of the self is left to others to decide. Is it any wonder then that the self that knows itself by what it is not is characteristically a self out of control, in that the assumed identity can be altered by the slightest gesture of independence on the part of the other? *Oh no! they are different than we imagined, and, therefore, so are we. . . .* For when one's non-being is conceived as an externality, one's capacity to change—to become otherwise than one is presently—is alienated also. Thus our barbarians become the free ones, though unfortunately, at this point, they can't know that for themselves, and so we feel obliged to know and act upon that freedom for them. In this way, we act as their orphaned consciousness, while they become models of our virtual freedom.

In loving relations there is a certain naturalness to one's being with others, an ease within which difference is brought in close, attached to one's identity in a reciprocal way, and adored. That ease of adoration which one experiences in love is perhaps an animal, perhaps an intellectual identity with others: knowing, really knowing that they are with you, that they can have your trust, and knowing that they trust this too. It is this kind of reciprocating surety that makes humans capable of a behavior like love that is at once so startling and so natural.

There is then a kind of likeness between one's comportment toward a lover and toward a barbarian, with the difference being the fulfillment offered in love, something that remains absent in the hatered toward strange other. This fulfillment begins with the acknowledgment that one needs the other, whereas in the formation of barbaric others a civilization's own lack is unselfconsciously projected, so one can never come to know the self's own absence in the other. The feeling of identity in hate thus remains an unspoken lacuna, an abyss for self-consciousness as black as the moon in shadow. Virilio marks the perceptions of the first astronauts on the moon, who noted a weird, bifurcated quality to the light, as if objects were turned either off or on by their illumination.[8] Something similar is true of the hate of barbarians: that there is no penumbra of shadow inflected into that perception of the other; they are known with a surety that only a digital imagination will allow. How then were there barbarians before the world was rendered into a series of zeros and ones? One can point back behind modernity to the roots of our word barbarian in the Greek *barbaros*, though by varying accounts this meant simply one whose language was not understood, whose voice was heard as so many analogue rhythms repeating the syllables, *bar bar bar*.[9] What we have in late modernity goes far beyond attempts to reconcile the barbarian with, for example, the language of Christianity, translating a *Gitchi Manitou* into a *Deus ex Mundi*, for presently identity is referred to in terms of a strict nominalism, as in the unblinking, unquestioning identity of a biometric identification card. Identity is something to be rendered from the subject, whether by chemistry or behavioralist psychology or special techniques; the identity of the barbarian now is something that would be found out from an aerial view, accessible to the algorithms of a killing machine that would make the only translation necessary from 0 and 1 to "them" and "us," with identity formation ultimately reduced to the identification of targets. All that remains to humanize the violence of a monitor refreshing is the cruel charm of ignorance—the dumb self-assurance that knows an other only as a nothing. These singularities can connect then only when the other is dead.[10]

So is love instead the answer then to the barbaric exclusion of identities? Perhaps this is a bit too much to ask. One would have to achieve some level of self-consciousness, knowing what one lacks first; and then, once that baseline level of critical consciousness is achieved, one could begin to recognize and then get to know that other person or people, perhaps seeing in their culture some practices and perceptions that compliment the hypertrophic developments of one's own culture. So, a technological civilization could learn from one sensitive to natural integrities; or a devoutly religious people could be made more whole through a measure of critical doubt. But these kinds of exchanges could succeed only if they were initiated in a spirit of listening and receiving; otherwise, the hated barbarian could be exchanged for the loving barbarian without really getting to know what love or hate or the other are about: Hate would be forgotten as violently as it arouses, love would become an emotional measure of ignorance, and the image of the other would remain as strange to "them" as it is fascinating to "us."

AN INTERRUPTION AND REASON FOR LOVE

> . . . the very fact of being is placed beyond the distinction between the perfect and the imperfect.[11]

Love for Levinas cues into an openness to being beyond its categorization; yet what would such an openness mean if it necessarily transcends the determinations that make our experiences communicable to others and intelligible to ourselves? If one were able to gain some kind of revelatory perspective on the basis of being from the perspective of being itself—prior to the differentiation of elements, the dispersal of cosmic unity, and the introduction of apparatuses of observation between world and observer—then something like the ontological speculations that follow might appear beyond appearances. Otherwise, and from a strictly human perspective, the following likely story is for those for whom the chaotic unity of love has an immediate resonance. For there are some things that can only be grasped on the basis of love.

Hypothetically, on the basis of a spontaneous intellectual affinity—that is, he likes the idea—Zizek posits that the basic substratum of reality is a void.[12]

Cosmologically this is an undifferentiated soup of unstructured being. Mythically, the ontological premise takes on the familiar form of an ocean of timeless becoming, as in the Judeo-Christian cosmogony, in which:

> . . . the earth was a formless void and darkness covered the face of the deep, while a wind from God swept over the face of the waters.[13]

There can be no privileged access to this kind of perspective on being at its formless origins. There is no standing back or seeing with perspective here; even God in this story is manifest not transcendentally (there being no determinations yet to transcend), but immanently as the first initiating ripple on dark waters. The all-involving inscrutability of an original union of order and chaos, along with the magical, cohesive formlessness of water that is so immediately familiar to all human beings, induce this archetypal image to be born forth in similar forms from out of the diverse legacies of the world's myths. This is the same blood-red, swimming blackness that we all see when we squeeze our eyelids shut, wherein the closing of the gates of reason reveals the origin of mythic archetypes in a pregnant abyss.

So the Finnish cosmogony in the *Kalevala* has it that,

> Ilmatar, daughter of the wind, lived long ages in sacred loneliness in the smooth and endless gardens of the air. At last she grew tired of her life in boundless space, and stepped down to the waves on the broad back of the open sea.[14]

In his frenetic telling of what is much the same story, Zizek's removes any cosmological priority given to the unity that pre-exists creation and instead attributes infinite value to the formlessness of the underlying flux. This flux, the apparent happenstance of the original moment of creation, is now considered for its ontological, epistemological, and even ethical values. Hence, a conscious embrace of the randomness of existence is projected as the proper comportment for human beings toward each other and as the most substantial way of knowing another. So when we take on relations to others of supreme ethical value—when we love another human being, for example—what we are doing, according to Zizek, is fixing upon another for no real reason. In his interpretation, the fixing upon is the real reason, and original ground of whatever reasons we may claim afterward. The reality, however, is that it is we who have been claimed by our reasons in our own original act of laying claim to love. It is thus that love comes before reason and acts as its ground.

Just what is it then that allows one to fix one's attention on another being and to take them in lovingly as a genuine object of care? In what is perhaps the most enduring philosophical treatment of love, Plato assumes (on the authority of the priestess Diotima) that this fixing upon was a natural, rational response to the universal forms of beauty or goodness, shared in by the beloved. Because those perfect forms endured, it was thought that one could not help but look at the beautiful soul and be struck into stillness by it. This attitude has become a kind of common sense, that we love others because they are good, or because of some beautiful aspect of their body, mind, or spirit.

So, contra common-sense, how can Zizek say that it is the imperfect that deserves our love? Or what is much the same, on what basis can Levinas insist that the fact of being is beyond perfection and imperfection? I think that, in terms of fixing one's attention on another, as in love, or of fixing upon a defining aspect of being, as in metaphysical thinking, what is common to both is the uncommonness of the loved one or of the "fact of being." Something catches one's eye as it courses over the smooth surface of phenomenal reality, in the field of which each being seems to blend into what is beside it simply by being looked at successively. For while there may be something lovely to an integrated field of appearances, smoothed over by the wandering eye, that kind of looking performs what is essentially a casual, confirming gesture, as in, *yes, that's perfect, and that too, and also that, and that. . . .* It is a machine-like looking, expressing the kind of appreciation that one might experience upon looking at a line-up of swimsuit models; but how one could fall in love if looking were always like that? It would be necessary for imperfection to break the frenzied trance of surveillance—so, a part of the person sticks out more or less than another, or a slight, sincere smirk breaks upon a practiced, cosmetically tuned smile. The possibility of love opens up precisely where formal perfection falls short, arresting the flow of appearances into singular moments of attention.

And why this and not that love? Here the sorts of stories that are told in knitting together a world of coherent, transparent identities—that *verum factum* world-view wherein one human-made phenomenon keys into the next—fall short of the interruption that love is. No well-adjusted person would fall deeply in love. Hence the inscrutability of love: It is always dependent on someone else's free choice, and there is neither externally verifiable obligation nor any rational capacity that can justify it, except perhaps after the fact. And then it is already too late for reason to be much more than a cover-up for love.

Being honest then means making these bonds of care in full acknowledgment of the fact that one loves some particular person for no pre-supposed reason—that is, because they had embodied a perfect form of beauty, or goodness, or etc. This is a way of breaking the spell of a Platonic idealization of love that would otherwise distract from the primal act of fixing upon another. In an honest act of love, the lovers consciously take on the chaotic quality of the reality that subsists underneath their choosing, so that instead of making statuary of beloveds, they are loved for the fact that they have the power to move. Is there some inherent capacity to do this that subsists in their bodies, minds and souls? It is impossible to say. This is an acknowledgement of Nietzsche's dictum from *Beyond Good and Evil*, where he writes that,

> For sheer love of humanity one occasionally embraces some random person (because one cannot embrace everyone). But this is the one thing one must not let the random person know.[15]

Love is in this sense a form of wonderful deception, of both ourselves and others, and it is in this sense that Zizek calls love evil. Love disrupts the amorphous soup of underlying drives by fixing upon one thing as the singular object of attention, upsetting the primordial unity by introducing the object of desire as a kind of aberration, loved for its imperfection. The imperfection itself is thus both of supreme importance and entirely incidental; or to repeat, there can be no reason for love. Love is de-linked from a rational object of desire, with what remains consisting simply of the desire itself, grounded in its own desirousness. What keeps it from narcissistic wasting is the act of looking into the eyes of a lover, truly in love, where the only mirror in the room is the blackness of the pupil of the other, within which is reflected an undetermined self. Hence the need to love, so as to see others as essentially free, and thus to free oneself also.

Or as Nietzsche writes:

> Ultimately one loves one's desire, not the desired object.[16]

All that saves the erotic will is the constitutive power of desire, wherein the loved one is not imagined into being (this would presuppose an image to begin with) but whose being is experienced as a positive absence, a rift in the train of correspondences that make the world appear to be of our own making. Thus all of the reciprocal actions by which one proposes this and that to the world and waits for it to say "Okay," start to seem like so much background static noise against which the signal, unconditioned response stands out like a meteor strike on a lonely planet.

If love is ultimately an irrational fixing upon another, can one love humanity as a whole? Should one try?

If one really loved another, so as to experience true union with them, then they would become as much of a question as the self is to self. The perfection that is in imperfection, or what is the same, that which is beyond perfection and imperfection, if it were incorporated into the core of self-knowledge, would undo every alterior-determination that self sets upon self. The world would glow with a mystery that would unravel every use, term and condition, every assumption and expectation, via which one identifies the world as one's own because it is *like this* . . . or *like that* . . . or *not like this* . . . or *not like that*. . . . With these alterior identifications undone, all that would remain of identity is the care that is experienced in awe, which overwhelms consideration of good or bad, and, being beyond interest, holds a being out over a world that loves them to death. This is the terror and the wonder of universal love.

A HANDBOOK OF INWARD CULTURE FOR OUTWARD BARBARIANS[17]

> And yet the value of European civilization consists incontestably in the aspirations of idealism, if not in its path: in its primary inspiration idealism seeks to surpass being. Every civilization that accepts being—with the tragic despair it contains and the crimes it justifies,—merits the name "barbarian."[18]

Who is the barbarian here, and what is it that made him so?

First, there is a certain despondency towards being as such, in its pure facticity and givenness. In response then, a realistic if fatal acceptance of being is given up precisely because of its determinations—and it is evident that there are many—as Levinas insists upon the transcendence of being by another name—something like the desire for "otherness" as such, though it could more simply be called desire.

Does such a translation of "alterity" into love answer Levinas's aim to describe transcendence in the face-to-face, accruing a discrete worldliness to some sense of being beyond being? Here what is barbaric is understood as an acceptance of being as a given, a comportment that goes along with a rigid assignment of definitive qualities to others; or put otherwise, barbarians are those who think they know some others who are the real barbarians. What saves this definition from its own circular logic is that one can leave it to others to decide whether they would be barbarians, simply by a gesture of calling others the same: that is, the definition is not a device for any diagnosis or rehabilitation of the barbaric other, but it can be employed only as a thought practice for self-knowledge. In terms of rehabilitation within the very real context of a global humanity, with collective fates now linked by technological and natural limits, what would save a people from their own barbarity is the capacity to recognize another culture beyond identifying with it, that is to recognize it in its differences. The question that then arises, and the reason for invoking love as a mediating term, is just how one would come to know difference beyond identity.

It may be helpful to pull back this ontological questioning to perhaps more familiar grounds of intercultural communication, in order to establish the value of love as a way toward identity in difference. Perhaps the most universal point of contact between cultures is the food we eat—human beings being the omnivorous and ingenious creatures that we are. This is certainly one of those determinations that defines what we are as a species, that we must eat, and so come to terms with the fixtures of necessity by embracing our fate and feeding ourselves—whether as individuals, family, friends or guests—with a mixture of enthusiasm, creativity, and those inherited recipes that tell apart one family of kitchens from another. Marx suggested that men make their own history, though not under conditions of their own choosing; here something similar could be said of the food we make and consume: that it is both a reflex of necessity, and a product of a generational fate, and at the same time an avenue for a novelty in combination that has not, at least to the present, exhausted the resources of our incredibly edible earth.

Thus, there is both significant variety in the foods we eat, and, because we have roughly the same sorts of bodies (acknowledging the subtly different digestive and metabolic capacities of different peoples, from life-long vegetarians to traditionally meat-eating First Nations) and because we sustain our bodies from an earth that is humanity's common heritage, these differences converge around certain common tropes as necessity is reconciled by resource. So, given the broad spans of time and geography intervening, one finds perogies in Eastern Europe and dumplings in Asia, rotis in the Caribbean and burritos in Mexico, soup and porridge everywhere, and smoked fish wherever there is water and a need for the preservation of food. The more basic the foodstuffs, the more their form tends to converge, for the same reason that humans walk in much the same way, one foot in front of another, out of this particular juncture of our similar bodies and the planet we walk upon.

Of course, there is a lot of variety in food, as various as local ingredients, histories, recipes, and the cooks who make our physical, cultural, and communal sustenance, but it is worth noting that this variety is greater in those foods that are made in a spirit of freedom and diminished when it comes to the staples—with the starches and the roasted meats at the end of necessity, and the desserts of the world representing pure resource and the freedom of food for the sake of food. And while one dumpling may taste as good as another—though different—the variety in the desserts and other treat foods of the world can easily outdo the flexibility of even the most cosmopolitan palette: like hot pepper candies, or gelatinous cold sweet bean soup, or pig's blood crepes. Stereotypes around what is decent food and what is disgusting are instruments of self-correction as much as they are instruments of domination over a stereotyped other. If "they" are one way—perceived as rude, dirty, or gross—then the "we" is imagined to be otherwise than that—advanced, hygienic, beautiful. Indeed, stereotypes may be effective only in the management of others—as in the dream of the other ordered and available as knowledge—when that other becomes oneself.

This is not to say that everyone's taste in porridge is the same—some like sorghum, others like oatmeal—or that all like their meat roasted to the same degree of doneness, but these are in any case all recognizable as common fare, though with different flavors.[19] But when a person is free from hunger to make up a food irrespective of the necessity of eating, then the divergences between peoples become most apparent. Peoples converge around necessity and diverge as resource satisfies those dictates, making themselves free from hunger, liberated to eat if only for the pleasure of eating, in an instance of freedom re-inscribing fate into the life of humankind.

What this means for intercultural discourse, and for the possibility of coming into a loving relation with those we don't know, is that the freedom that all people love, in the very basic sense that none would choose to be compelled by hunger, must always remain an open signifier. That is, one cannot know what the freedom of another means definitively any more than one could predict the desserts of distant cultures, as if one could pattern the specific desires of others. In the most profound sense, the freedom of one's own self-determination must be at least episodically refreshed if the culture that contextualizes its own ideas of freedom and purpose is to sustain itself in a set of common aspirations. This clearly does not necessarily offer a cross-cultural translation into a common longing for progress in history, whether the determining goal would be a lockstep liberation toward liberal markets for investment capital, the exponential increase in the speed of computer processors, or the transformation of nature into plastic stuff marshaled for human use. There is no definitive sense in which the moments of liberation from necessity that are opened up by human resourcefulness are reducible to any core historical drive, whether it is considered as a reflex of human nature, as in the Hobbesian concept of progress motivated by unlimited acquisitiveness liberated from chance by the protection of a central security state, or as an immanent unfolding of a distinctly human capacity to negate the given, as in Hegelian-derived theories of progress.[20] Different political formations remain possible, various historical tangents occasioned by crises and the unexpected outcomes of complex systems are likely, and the novelties in combination that occur when the creative energies of different people come into contact ensure that human-made forms will be almost as surprising as nature is in its various and changing formations. Wherever there is this wonder at the precarious potential of human beings to set new goals for themselves, there is the perennial possibility of political life, of really human life, and then also, subsequently of philosophy. For wherever two or more are gathered, one can never be entirely certain what their deliberative reason or spontaneous agreement will decide. To the extent then that this political capacity remains deeply inscribed in the condition of plurality that characterizes the human condition, dispersed as we are over the surface of the planet in uneven clusters, each drawn together by various kinds of love and hate, any claims to definitively project what others would imagine in their ideas of freedom would have to be prudently foreclosed, to the end of respecting that source both of differences, which together inform the capacity to change one's mind, the later being the particularly human freedom: that is to think otherwise.

LOVE AND DUTY

> In love, the one individual has consciousness of himself in the consciousness of the other; he lives selflessly. In this mutual self-renunciation each gains the life of the other, as well as his own which is one with the other.[21]

The specific kind of love experienced in family life is Hegel's point of departure for a conception of freedom grounded in duty, toward a civil religion that would fill out a civic sense of freedom from the emptiness of negation toward positive determinations of self in freely formed, autonomous associations wherein there is a duty to others. For beyond the enclosure of the family, the projection of love in freely formed terms is civic friendship, where by mutual obligation each becomes the self in the other.

Why is there the constant temptation to include the definite article in front of the singular Other? Is this some legacy of metaphysics working past its end even in the critique of a logic of singularities, of identities formed according to the logic of I=I, then unformed according to the crude inversion of the same, with "the" Other standing in for "the" self in that equation? Are both "the other" and "the self" somehow suspect as concept formations?

Irigaray for one suggests that something of this kind has been the case: that a claim to unity has inspired Western thought from its origins to the present, and that any inversion of the sameness of the same into the otherness of the other is moved by the same animus. What remains for her is simply to think difference and to find a basis for love in that. Beyond emphasizing the limits of any attempt to draw reality together with a whole web of "equals" and "does not equal" signs, the hope is to come to know a self that is essentially mediated, that finds its own being in difference. This kind of identity formed out of its own being different and held out toward another for a sense of difference would be a self borne out of relations with others, and a person who can become other than who they are, not by forming an identity with some other, such that the other becomes a mere subset of the self, an identity in likeness, as in *Look, we both like beer . . . . We may be different politically, socially, economically—but here is a cultural sameness—we belong to the universal class of beer drinkers . . .* but rather in the formation of a mediated identity developed out of difference. This developing awareness of self would begin not with the habitual presumption to know the other but with an inversion of one's own identity, such that we become strange to ourselves and begin to know both the stranger and ourselves through this existential unease. The ethic here then becomes one of hospitality, with the true foundation of culture being how to make ourselves at home in a world that does not tell us who we are. We can only ask this of ourselves and others: that we learn to dwell in the care of others.

One of Irigaray's recurrent points is that this kind of mediated or relational understanding of identity, if not characteristic of feminine nature as such, would represent a progressive enculturation of the female sex. So if masculine identity has evolved within the course of Western thought as a specific cultural development, with priority given to the I as somehow immediately disclosed—say in the individuated soul substance of the *psyche*, or the atomistic liberal self—then a full recognition of the divided character of human nature—at a biological level between male and female—and the creation of a properly mixed-sex politics require an elaboration on this mediated quality of identity.

Since such an identity would have to move between the giveness of sexual differences and the cumulative meaning attached to those distinctions—that is the stories that we tell about our differences—some narrative illustration would be helpful. There is an enchanting example of the creative tension that exists between the sexes in Shakespeare's *Midsummer-Night's Dream*, a play of misplaced, fixated loves, where Oberon and Titania have out a domestic dispute in a forest clearing by moonlight.

Ostensibly, the cause of the row is a dispute over who should keep a young changeling boy desired by both, but as in many domestics, this original cause of the fight and their separation spins off into a litany of accusations. In such a spirit of spiraling disputes, the goddess Queen Titania accuses her estranged husband of conspicuous infidelity, charging:

> . . . but I know
> When thou hast stolen away from fairy land,
> And in the shapes of Corin sat all day,
> Playing on pipes of corn, and versing love
> To amorous Phillida. Why art thou here,
> Come from the farthest steep of India?
> Your buskin'd mistress and your warrior love,
> To Theseus must be wedded, and you come
> To give their bed joy and prosperity.[22]

In an interesting deflection from her jealousy at their own lapse in love, Titania accuses the philandering Oberon of denying these others' husbands the "joy and prosperity" of their marriages. In this she seems as concerned with the health of those other loves as with her own. So one breach of sanctity leads to another, as another other gets in between and fixated upon as intervening subject in the communion that love is. If identity is mediated, then the desires of others would be inflected into an ethic of difference, acting so as to consider the other before oneself.

FROM FAMILIAL LOVE TO THE FEAR OF OTHERS

The mediated quality of desire is essential to its being, with the thing itself being predicated on the presence of another. One could not love in a void (if it were pure even of history), with its patterns of gain pointing the self outward to others, or as Hegel describes, "each gains the life of the other, as well as his own which is one with the other." Beyond ideological interpretations, this is no simple matter of one person's self-abnegation or the suppression of individuality (at least so long as one has the emotional intelligence to distinguish civic spirit from loyalty to family). For what is the family that Hegel is describing here in the absence of the distinctive presence of any one of its members? It seems that what remains of love within the family is just that capacity to remember itself. Both the life of the family and its consciousness consist in its memory, whether genealogical and genetic, or personal and psychological, or social and historical. There is really little else to its identity than these patterns of recollection among those who love or have loved each other by direct acquaintance. Love, after all, needs that locale of affinity experienced in the closeness to others and in the naturalness of the experience, as if humans were ontologically suited to holding themselves out to another. Thus,

> Parents love their children; they are formed by nature to do that. But it helps when the object of their affection is so eminently lovable.

Every such remembrance of a loved one owes the limits of mortal memory:

> ". . . an intimation of true immortality for we of mortal flesh."[23]

The words of a father bereft move a person to reflect on the constitutive power of remembering a loved one, both for the love itself and for the loved one remembered. There is a kind of spiritual substance in the absence that makes memory of another possible, with the loved one called into presence whenever a story is told about them. It is in this sense that love remains even in apartness, in the absence of an immediate presence, and even sometimes after it has ended. If there is a trust in humanity expressed here it comes back to one's own self image, consisting in the hope that prior loves would remain intact, even if only as memories of another, preserved in that precise sentiment that called them out of the stream of incidental encounters. If a love is lost and the feeling retrospectively turned into something else, one risks losing all that surrounds that time and place as constitutive of the memories that gather the self together and give it its sense of being a whole. If love begins with the experience of being dislodged from the stream of incidental encounters—the *and this and then this*, the *hellos* and the *how are you*'s—

such that a truly startling revelation of another occurs, then the memory of love too would have that same interruptive character, acting not to consolidate but rather to break off the mechanical succession of events. Thus, every memory of a monumental order would in measure surprise, disclosing the love by calling us back to an original break—like the first look or the first goodbye, or so on: Pick any origin, like a first star in the growing dark of a new night sky. The memory itself is a kind of act, interrupting the apparent immediacy of sense perceptions, and re-initiating—every time it is enacted in a memory relayed—relations to others, both to the distant and the departed. The self thus remembers itself in relation to others in every love remembered.

Short of an emotional lobotomy, being of good faith toward one's time-distant loves, attachments, and judgments adds an integrity to the self that can keep itself together even as its loves find their own constitutive relations. The hope then is that we would remember our loves as they would have themselves remembered. Culturally speaking, there is much to be learned from the mythic traditions here, in a collective, civilizational memory by which we go about loving our own. What is presupposed by a hero, like a Penelope or an Odysseus, is that the self is occasioned by and coincides with its story. Every memory of a person is a story, whether it is the story of what one did today, told within the love of family life at the dinner table, or the story of a life recalled in the absence of a loved one, when they are brought to presence in private recollection. The self stands out as a self precisely because of the quality of difference distinguishing him or her. Regardless of the scale of recollection, what matters in this is not the consistency of memory (i.e., there are many ways to tell a story, and for this reason identities are fixed only when they are decisively forgotten), but what is significant rather is the memory's holding out toward the interruption that every love represents, dislodging the self into constitutive relations with others. This is how precarious love always is, for to be in love is to have one's being owed to others.

Machiavelli was right in noting this aspect of love, namely that it puts one's fortunes at the inclination of others and is, therefore, always somewhat out of control. In love one is moved by a profound awareness that one's being is beyond oneself, or what is the same, that there is an aspect of another in oneself (at least if the love is reciprocal). Thus, being a whole person requires both projecting onto others in whom one sees essential qualities that are beyond immediate self-sufficiency—that is, to have the experience of need and desire—as well as opening up a space within, a little lacuna at the core of the self that others are free to fill in themselves, such that one opens oneself up to the dangers of love. However, if the aim rather is for complete self-sufficiency and an ontologically in-hand identity, where the self has all that it needs and defines precisely what its self-image may be (as in kind of xenophobic, militant survivalist image of self-sufficiency), then love is an evidently unstable basis for the creation of identities and for the formation of some sense of understanding toward the perceptions of others. Hence Machiavelli's halfway grimacing advice that it is better to be feared than loved, because fear is manageable whereas love is fragile, handing over part of one's being to the perceptions of others.

Thus, in a radical flip-flop of epistemological decision the famous exchange of love for fear would constitute both an entirely new beginning for political philosophy and the start not incidentally of the first truly modern political formation, the state, which, unlike medieval or classical communities based on faith or friendship, would have its basis instead in an outward and obvious sensibility—fear—that could be known in a way that a god or a friend or nature could not: tested, measured, adjusted, and dialed in to create a direct circuit between centers of power and their application onto individual psyches and social behaviors. Whereas in love even the boundary of the body and the outerness of others are ultimately dissolved as an entering into absence with another, in fear rather both the object of fear and fear's object come into crystalline distinctness, in an awareness of the potential source of pain and where its application will come from, reciprocal and rational as if by engineered design.

If the aim of modern philosophy is to bring, in Descartes's words, clarity and distinctness to knowledge, then something similar can be said of modern politics, which aims to take those perennial and unanswerable questions of "the Good, the True, and the Beautiful" and the attendant, tenuous politics of faith and friendship and resolve the incommensurability of political life into a science for the management of human behaviors. The first step towards this end in the early moderns is to simplify human natures, with all of their complex longings and transcendence-prone short circuits of the imagination reduced to the level of simple behaviors, and this means first resolving them into a lowest common denominator of motivations and from there reconstructing a social psychology out of those atomized appetites and aversions. What remains are individuals who know no better than to care only for themselves: those who are lacking in the capacity to imagine a fulfillment in the indirect, ricocheting experience of others; those who are used to living inside the margins instead of at them; and those content with the offer of security for personal indulgence in exchange for the evacuation of the commons and the centralization of political authority in the state. Yet, if this solves the technical problems of politics via their reduction to issues of corporate security and private accumulation, then it does so at the expense of making automata of human beings, who are understandable and manageable then only to the extent that they think like machines. Whoever does not, invoking love or faith or friendship in public stands out rather conspicuously as a factionalist, or a zealot, or a revolutionary out of season.

And yet, this is also one of the great paradoxes of modern political formations, that although they begin with and return to the atomistic individual as the locus of value and source of political legitimacy, there is also the perennial attraction of national or sub-national identities, formed within the artificial borders of the security state. There are too many reasons for the resurgence of non-atomistic identities to explore in depth here. In part, it is in response to the hollowness of the security state and, if one is allowed some classical romanticism, out of a natural yearning for the community of others. From Aristotle's *zoon politikon* to Hegel's *Volkish* national community, there is a persistent sense that even if freedom is the ultimate goal of human striving, a freedom that is experienced in solitude, however absolute, cannot be a fully satisfying one. The question of what is worth doing with our freedom needs some public forum in which to be asked and some means of inter-generational collaboration to give it substance, at least if freedom is to be understood as more than the freedom of license or, as Machiavelli describes, the simple desire to be left alone.

For Machiavelli, locating the agent of freedom as such depends on the gradual exception of certain social anachronisms, like aristocratic nobility, which was in his time just then on its way out, with a new form of republicanism struggling to cross over the boundary of the ancient and modern traditions. If the desire to lord over others was to become suspect, it would have to be somehow excluded from the common longings that hold any society together. Machiavelli's rhetorical strategy for doing just this was to point out that the drive to be more than common was an impossible desire, tied up with the competing drives of other would-be aristocrats. Perhaps Machiavelli does not properly register the significance of political friendships in balancing these kinds of competing drives and sees aristocracy rather like the ethic of wolves queuing up for their share of a kill; but in any case, whatever the fragility of aristocratic egos, compared with the tenuous vicissitudes of parity among unequals, he thought the simple, self-assured desire to enjoy one's own freedom to be a more solid basis for a social order. And in his glancing advice to princes, he urges that even when they do gain power, it is in the people that they can trust and upon whom they must come to depend. Hence the need for a social psychology suited to states and attuned not to points of distinction but rather to questions of likeness.

Hence also the deployment of barbarism as a tactic to frighten and mobilize a nation into being. Machiavelli's *Prince* shifts gears rather rapidly in its final chapter, "An exhortation to liberate Italy and free her from the barbarians," with the technical modes of statecraft, spare historical examples devoid of moralism and detached perspective that characterize the twenty-five previous chapters radically altered in tone and substance. If the rest of book is a neutral tool, the exhortation is most assuredly about what ought to be done, snarling,

> This barbarian occupation stinks in all our nostrils![24]

Here the technical advice given throughout the Prince is animated by a rhetorical flourish, augmenting statecraft with poetic art. For what can be the basis for this new, expanded "natural" community, the nation? It is not actual associations between citizens, as in the Aristotelian ideal of the polis, since the nation-state is simply too large for all citizens to know each other or to be addressed together; nor can citizenship within a nation-state be based on actual friendship. The nation-state's sense of natural community is in this sense more amorphous, more poetic, more along the lines of an "imagined community."[25] In Machiavelli's vision, this kind of belonging is held together through the remembrance of military history as narrative for mobilization. Here the memory of war is tuned to maintaining war as a capacity of the state, enlisting what remains of civic spiritedness contra the "barbarian occupation," making use of our barbarians to confirm an identity in opposition to them.

NEGATION AND SELF-KNOWLEDGE

> In a world where protagonists rush into the abyss of total degradation before exterminating themselves physically, new types of hostility are bound to emerge. Hostility will become so terrible that may be it will not even be possible anymore to speak of enmity or hostility. Both will be outlawed and condemned in due form before the start of the operation of extermination. This operation will then be totally abstract and absolute. . . . The negation of real hostility will in this way open the way to the work of the extermination of an absolute hostility.[26]

Isn't this radicalization of differences precisely the endgame of politics that Hegel's "negation of a negation" gets us to, where freedom confronts its own excess, and imagines succeeding against it(self) through yet another excess and radicalization of the underlying process of identity formation by elimination? Thus, in the hope of attaining a kind of ersatz transcendence, a singular spirit and common drive is read into every preceding historical moment and all other peoples, with progress toward universal civilization serving as secular religion, with a direct, psychological connection made between individual and humankind. Practically considered, it was simply a matter of finding a certain enough emotion (as if the psyche could be reduced to some semblance of a single feeling immediately apparent to all)—one that would decisively undo the seething manifold of considerations of the good, with these put back behind modernity as too unsettled a cause for efficient politics.

There is a rude truth in Machiavelli's judgment that "to be feared is much safer than to be loved," if only because fear can invent its object and so know it just by the ontological security of a *verum factum*. And yet still, what is ignored within the self-made truth is an apparent absence of desire: It is a truth that cannot possibly love itself, and since all truths are inherently jealous, it is a truth that can not know itself then either. Or if it does have a reflexive capacity, it is truth that knows itself merely as fact; if that kind of reflexivity within the idea itself (like the ancients' concept of *nous* as both the knowing and what is known) is even still possible, living as we do in the shadow of modernity and its objectification of truth, to indicate that the truth itself is capable of self-knowledge. *Or else,* modernity might threaten, *or else, your truth will be my truth. All truths must identify.* Love is thus set aside in favor of a predictable identity fixed absolutely in fear, politically and personally instrumental as global humanity's reactive social bond.

Modernity's dominant identity has been transparent for quite a while now, at least since transparency has stood in as discursive truth in the liberal tradition gone global. If the question of identity is exhausted, largely reduced to issues pertaining to adequacy of representation (except in those rare instances where identity is known by difference), then the political question *sine qua non* can be traced back to the ontological question of what is difference, or put somewhat more speculatively, what is the interval of transcendence (such as experienced in love, or religion, or cultural ecstasy—wherever one identifies with difference) that is opened up by the presence of others?

There is difference in this sense within every identity, and the questions one asks of others are precisely those that lie unanswered in self-consciousness. It is in this that the drive for self-knowledge is initiated not by intention, but by an absence within that calls a person forth toward difference; i.e. toward some completion of the sense of absence in identity with another. Thus the basic philosophical question of *Who am I?* folds back upon the fundamental political question of *Who are you?*

ON LOVE LOST AND GETTING IT BACK

What does love have to do with politics? To put love in the context of other political emotions, it is remarkable that it is taken as a common-sense given that fear be treated as the inherently political emotion *sine qua non*, whereas an association of love and politics requires some convincing. So in Hobbes, for example, the fact of a fear of violent death is given as both axiomatic, and as a universal emotion upon which the modern state—our ersatz political community—may be founded. Hobbes's obsession with fear may be a bit of an exaggeration and hypertrophy of one isolated aspect of the human condition, and in a sense also a slip of logic, in that, given the equivocation that Hobbes has toward the coherence of the passions in general, there can be for him no basis of similitude between the various objects of the passions. That is, according to his political psychology, what is certain is that there is an underlying set of drives—appetites and aversions—though we cannot agree on just what our appetites may be for, or what overlap there may be between desires. Thus, there is, in this formative early modern account, apparently no *summum bonum* that we can love in common, nor any reason encoded in our desires that would lead to a harmonious arrangement for either each of us as individuals, each with our own wonderfully complex messes of competing drives, or of our encounters with difference in the presence of others.

In an interesting turn of liberal logic, it is precisely the idea of a common good desired by all that for the early moderns supposedly brings the worst kinds of violence upon political communities. It would be worst in the sense of a totalizing violence, an ecstatic, ritual politics enervated by a sense of limitlessness combined with extreme inwardness, one that cannot recognize difference because its own good is total. The turn initiated by these classical liberal thinkers is that notions of a common good dangerously infect public discourse by inspiring a fervent passion into politics that is better left to private hobbies.[27] So for Hobbes (and for Locke also) one can be passionate about what is private and inherently one's own because one safely possesses it. One is passionate about property in the same way that one is passionate about model train building, or sailing, or needlework. What makes these sorts of private hobbies acceptable is that no one else really cares about them; your property and your toys are your own. If those sorts of things move people, they do so in ways that are unlikely to lead to collective political action. So we are left with what has been called a nation of shopkeepers,[28] the passions of each bled off by trivial pursuits, each minding their own business and uncaring for others unless those others somehow interfere with one's own private business (though properly speaking, business today on the whole is corporate, monopolistic, and not private). This reflects Hobbes's idea of a common good—that is, a community with an absence of a common good, of politics without passion—what could be called a politics of safety first. This derivation of political community from fear is thus invoked to justify the foundational responsibility of the modern state—ensuring the security of its citizens and their property.

This legitimating role of state sovereignty resonates with subjects, Hobbes argues, irrespective of our cares for what is good, because we are all metaphorically hard-wired to fear our own violent deaths. There is, however, an apparent leap in Hobbes's logic here: that is, why single out the fear of death as a shared object of passion? Clearly not all are equally afraid of death, notably the religious, the philosophic, martyrs and saints, or simply those who are brave or otherwise worriless about their own personal demise, perhaps being vested in a care for a good more substantial and demanding an imperative than one's own self interest could be.

The slightly paranoiac, twisted sense of Hobbes's axiom, singling out fear of a violent death as *the* animus of political reason is made possible by a curious turn of liberal logic: that is, those who do not agree with the proposition that self-interest defines human behavior are precluded from civil reason and considered either mad or criminal. It is an instrumental logic in the sense that it can be easily applied, that is, by locking up the insane (or if they are of a milder sort of mad, otherwise confining and directing them through institutions) and killing the vainglorious criminal or martyr. To apply a biological metaphor, by demonstrating those kinds of behaviors, the outliers have shown themselves to be hardly human (after the model of *homo oeconomicus*—the calculating consumer) and so are fit to be isolated from the body politic as one would a virus. Hence the extrajudicial treatment of these remnants of the pre-political: Being beyond the limits of liberal civilization, they are the excluded and nothing more.[29] For in the absence of submission to a logic that identifies reason and self-interested fear, the security state has no psychological link to those unafraid of death, since this fear is taken as the ultimate basis of civil relations. By resisting the state with their very lives, the excluded are thus effectively returned to a pre-political condition, expelled from the state's protections and provisions and exposed to the raw basis of its power; they are each and every one microcosms of that whole, sharing in the nihilistic freedom of an unrequited death, "pure negation entirely devoid of mediation," a death that cannot be returned and exchanged for justice, "a death that achieves nothing, embraces nothing within its grasp; for what is negated is the unachieved, unfulfilled punctual entity of the absolutely free self."[30] For in the absence of a properly civil culture (which Hegel for one sees as the synthesis between absolute freedom and terror), the mediation between state and individual must somehow reduce to a discrete drive, a kind of servomechanism of stimulus and behavior standing in for understanding. What is missing is a mode of reconciliation that need not kill what it cannot control; otherwise, the search for identity plays out as the redaction of differences within a self-referential logic that does not know its own limits.

There is a kind of self-confirming circularity to this applied pattern of thought, as in the brash imperative: Fear violent death or die. This is in the end Hobbes's effective response to the vainglorious, the courageous, and the martyr. For those who do not fear the sovereign's mace, or the symbolic presentation of the authority of the state, the common sense reasoning by which the citizens are deterred from crime and convinced to surrender total freedom, there is in the other hand the sword. Like most circular arguments, because it does not reach beyond the immediately evident, any intimation of transcendence is suppressed, and the argument requires the evidence of brute force to make it real. Yet whatever the argument's means of demonstration, if rigorously applied, Hobbes's political psychology leads to relativism: It is only anchored in a *summum malum* by a political formation that would make our fears real; that is, in the creation of the state as a monopoly on the use of organized violence untethered from any conception of the common good, with the desires for any such end sequestered to the private realm.

I would suggest that it is the legacy of Hobbes, of his instrumental, liberal form of reason and of the state that would embody that particular code that have contributed to the striking absence of love as a political phenomenon and as a question for political theory. The priority given to fear over love as the foundational political emotion is not self-evidently apparent to reason, but depends in good measure on the capacity to enforce that disposition to fear, through a combination of punishment and propaganda—the mace and the sword that grace the cover of the *Leviathan*. Then there is also Machiavelli's point that whereas those overwhelming passions of love and hate depend on an unsteady relation to others, fear can be controlled. This manipulable quality—as in, for example, color-coded terror alerts—allows for a much more managed form of politics than do love and recognition, which always depend upon others who are beyond strict control

In order to go beyond this preliminary critique of the Hobbesian state and the politics of fear, I would like to cue in on this essential feature of love and recognition—the experience of difference—so as to answer the concern raised by Machiavelli and Hobbes that love is too incendiary an element on which to found a political community, or as in certain conservative critiques that a loving politics would be an ideological, totalitarian politics, within which those who do not share in that common love would find no place for themselves in the the company of zealots.[31] This notion can be sustained by an exaggerated conception of love, portrayed as an embrace so perfect it suffocates, like the punishment meted out to the adulterous Aphrodite and Ares, caught in the act of love by Hephaestus and bound up together in a metal web, pressed together and kept from the pleasure of making space in between their bodies.[32] This would be the self-referential love of a Dr. Frankenstein, who desired an offspring that would owe its being entirely to him[33]; or like the witch in the animated film *Coraline*, who engineers an absolutely perfect world to lure in a child who she would turn into her puppet, replacing her eyes with sewn-on buttons, with no gap between the world as the child sees it and the way she wishes it were. Or from philosophy, this is akin to the *amour de soi*, or primitive love of self, extended to a projected identity with others and the world (as distinguished from an encultured love or *amour proper* that discerns based on distinctions among people and between man and nature).[34] This is the kind of identification that holds together Rousseau's general will, wherein the slightest indication of a particular, divergent opinion justifies a hermetically sealed statement of radical liberal logic, namely that divergences of opinion are vestiges of a backward consciousness, and that those who suffer different opinions can be "forced to be free" as the bargain of participating in a singular, unifying will.[35] In all of these instances, what is called love is really a projection of some apparent immediacy of self-knowledge, a solipsistic identity of I=I, onto others, so that others are lovable, in the Rousseauian sense, because they bear an immediate resemblance to oneself.

To which I answer that that sort of concept of recognition as projection of self onto other cannot account for the experience of love, because it lacks the capacity to move. One of the elementary aspects of love is that it moves us; as Derrida recites from common knowledge, "love moves the heart," or more precisely, love can divide the heart, tracing out a separation between what is desired and what is.[36]

The reluctance to philosophize about love that Derrida manages to maintain for a bit evinces a sense that the topic of discussion is too close to the mode of discussion, that the identity of form and substance in the incomplete question, "Talk about love . . ." would make any dialectic impossible due to lack of differentiation, as if the desire to do so would illuminate the desire to do so. Rather, in the sense of divided being that it expresses, love and the capacity for recognition call the self beyond the self into what are essentially needful relations with others.

Certainly, the paranoid meglomaniac who cannot bear the world as it is, with its manifest injustice, ugliness, and coldness could be said to possess a strong desire to eliminate that gap between wish and fulfillment, imagining a totalitarian perfection though that kind of perverted desire, turned in on itself like an ingrown toenail, that is in reality the desire for an absence of desire. What redemption remains for the ideologue consists either in a monkish retreat from the world into the perfect order of the cloister, the annihilation of desire via an inward turn of self-abnegation, or alternatively, a turn toward the annihilation of the conditions that must pertain in order for there to be love:

1. There is a sense of incompleteness (i.e., a person is not a whole unto themselves).
2. There is the presence of essential relations (i.e., one owes one's being to others).
3. There is a free relation to difference (i.e., a free being can become other than it is).

Love is predicated on a condition of incompleteness, for otherwise we wouldn't desire what is entirely of ourselves. After the manner of Socrates's mythologizing in Plato's *Symposium*, where the god Eros is conceived as the metaphoric love child of resource and need, love does indeed begin in part with need, but it must transcend bare need in order to find a measure of fulfillment through connection with others. This is where the resourcefulness of love comes in, registering the fullness of erotic experience from out of a moving feeling of lack and longing.

One may say that one is "in love," or that another is "in love," as if love were a state of being, a kind of sweet, immediate presence that embraces lover and beloved, but we tend to reserve that kind of grammatical usage for those who have just fallen into love, as if they have temporarily stepped out of the stream of events into some kind of original love, as in a honeymoon paradise state of suspended time. One may look at an elderly couple who are still obviously affectionate toward and doting on each other and remark that it is wonderful that they still love each other, but it is obvious at this point that love has shifted somewhere along the way from a static state of being into its verb form, that is, the loving elderly couple love each other in the verb sense of "to love," though having learned though time that time is not suspended nor distance removed though the act of loving, having felt the on-goingness of love over time and the durations of apartness that punctuate it as being most definitely a temporal, and therefore partial phenomenon. Love is on the way or incomplete in the sense that it registers the need for another who is not oneself, whereby the fact that we are capable of love indicates that we are not whole, perfect beings without the presence of some kind of endeared difference, saved thus by the presence of others.

EROS AS SEARCH FOR THE OTHER HALF

In Plato's *Symposium* on Eros, the comedian Aristophanes's speech is introduced as being different from the one that went before—what was a clinical account of love by the doctor Eryximachus—in that it shares in the emotional connection that belongs to the experience of *eros*. The speech is presented not at all from an inhuman, god-like perspective, but in a way rather specific to human beings.[37] Its humane sense of presentation manifests in its outwardly dramatic form—though it is difficult to say whether its intended effect is to be comic or tragic.

In this suggestive, if ambiguous connotation, Aristophanes' half-joking, half-serious speech is like *eros* itself, "where it is clear that the soul is powerless to say what it wants, but divines what it wishes and speaks in riddles."[38] There is in this sense a kind of parity between the form of presentation and what it would represent, so as to welcome the god of love into discourse between human beings through this likeness of form and substance. This would be a ground for "getting" others in their differences, a ground opened up by the feelings of laughter or of tears that put the intentionality of the world into three-dimensional relief, mapped out by the events that surprise the senses and condition the general sense of the "out-thereness" of phenomena, while reconciling distance through emotional response.

If recognition is a pre-condition for love, then a certain element of drama is a necessary term for both love and recognition, infused into everyday situations through the mutual stakes that are raised by seeing value in others and their perceptions, recognizing those others as essential to one's being.

DIVIDED LOVE, THE BROKEN CIRCLE, AND ORIGINS OF WORLD ALIENATION

The story of the circle people from Aristophanes's speech has become a familiar one, told and retold through the medium of narrative cultures so as to seem almost like an artifact of human nature and necessary historical outgrowth of the human experience. It is in this sense a myth that had to be told, as if the dramatist touches on an archetypal mode of association—between love, the circle, and the presence of others.

In his praise to the god of Eros, Aristophanes tells the story of these circle people, original beings who were in a way perfections of humanity, in the sense of being without any lack. He describes these proto-humans in the form of two bodies fused into one, with each bonded to their other as their other half, quite literally conjoined into an assemblage of four legs and four arms on which the circle people would roll along when they moved, circus trick-like. As two into one there would be no gap between them, no separation of bodies, being constituted into a singular subject. Eros thus perfected is not really erotic as such, offering rather a picture of union so total and transparent that it undoes the reasons for wanting love in the first place.

Aristophanes's story begins with the proposition that from these origins human nature has changed, that our originally whole nature was cut in two by the anxious gods as punishment for the hubris of a too-perfect humankind. This was the original context of hubris, as in a kind of overweening attitude and failure of recognition that would have humans assaulting the heavens. It is this trespass in the story that love would have to make up for and heal, bringing partial union with others upon whom we are dependent. For even if the gods have cut human beings in two, inscribing dependency into our very being, desire offers a way out of whatever world alienation is environ to the human species, so as "to make one out of two and to heal human nature."[39]

Love here becomes a call to a lost unity and a genetic reminder of origins, wherein the highest love is that which is most particularly our own in the sense of being complimentary to our divided nature, with what is our own being our own essential otherness. One must wonder then about the ends of love what is hinted at in its consummation. By way of a suggestive hypothetical, Aristophanes offers a kind of mythic thought experiment where he has the smith-god Hephaestus asking two lovers whether they would want to be fused into one, flesh melted together like molten metal, so as to enjoy a singular unity. Aristophanes surmises that they would so desire; reversing his initial ambiguity about the soul's powerlessness to name what it really wants, in the end he is unequivocal, saying "Not one on hearing this, we are sure, would demur to it or would be found wishing for anything else: each would unreservedly deem that he had been offered just what he was yearning for all the time, namely, to be joined and fused with his beloved that the two might be made one."[40]

This is a romantic notion, if expressing a certain kind of romance that sees identity as its aim. There is an alternative case to be made for the recognition of difference as an end of love. Indeed, Aristophanes cribs part of his story of the search for the other half from Homer, who relates the same dramatic encounter between Hephaestus, the patron god of techné, and eros. In Homer's telling, Hephaestus finds his wife Aphrodite embraced by the god of war, to which he responds by snaring the two together in a web of metal, binding them together so that they cannot separate, foiled in the act of love by too tight a bond, and shamed before the presence of the other gods.[41] This version of the story, quite differently than Aristophanes's, puts forward the perfect union as being its own opposite, in the sense of a perfection that undoes itself precisely through its own radicalization. The perfect, singular union of love really does put an end to it, with that inseparability making the union less erotic than it is suffocating, and more violent than loving. Perhaps this is to be expected. After all it is the god of the hammer and the fire who puts forward the possibility in both stories—whether it be offered as reward or as punishment—of a singular identity as a potential end of humankind. What remains of love in its offing becomes thus a matter of precarious survival, a sense of furtive connection that outdoes any technological perfection or shoddy fix of humanity into a species that would know only itself, in a 1=1 identitythat would undo the possibility of difference.

LOVE IS RELATIONAL

In love we realize that our being is dependent on others, that we are not essentially atomistic individuals bumping around in a billiard ball universe.

For while there are essential differences that exist between us, there are divisions of being within each of us also. Luce Irigaray makes this point very clearly, being insistent on this critique of the Hobbesian self and the liberal logic of self-interested calculation that accompanies it. This isolation of the self within a singular self-interest, reducible to a universal desire conceived as natural drive, is for her an abstraction. Rather, neither human nature nor the person are singular.

According to this view, human nature is two, man and woman,[42] with it being this self-division of spirit that for her drives historical progress and the unfolding of the species. This is our elemental experience of otherness, and it is not an otherness that is entirely alien to our identity, but rather, we are held out toward this other in an essential need. At first this need is expressed as the necessary condition for our being—that is, no one of us would be what we are if we were not first the son or the daughter of so and so . . . , and yet, who we are (distinguishing the who from the what) is not reducible to that genealogy. As expressions of a love that is not *of* some presumed other—as if they were a possessed object of desire—but rather *toward* them as a subject possessing agency and their own differences within, we become responsible for engendering a new culture of humanity through our loving relations with others, making difference part of our identity. Loving relations are in this sense constitutive of the subject as such, which "divines what it wants" and makes itself what it is through a love that situates a person as a node within a complex web of interrelations. A person thus exists not in some interiorized soul substance and hard kernel of subjectivity but rather in between themselves and others. One could extend this sort of conception to the private personality too; that is, a person is who they are for themselves. This division of spirit between self and other extends to the individual themselves, finding oneself in one's relations to oneself. One could address this in terms of self-image, where the self sees itself as if from the eyes of another, as in Rousseau's *amour proper*, or in more philosophical terms, as the basis of possibility of reflection, with a certain bending back of self upon self, not simply to see who one is, but to constitute one's identity through the reflexive act.

This is what Plato called *dialegein eme ematow*, to dialogue between me and myself. For one can question oneself—how is this possible? One can introduce alternative viewpoints, alternating *on the one hand, on the other hand*. One can read a Platonic dialogue, recognizing the particularities of the various interlocutors and bringing this variety of perspectives into one's own thought. Even thought in terms of its interiority, the identity of a person is always on the way and divided from within, just as the formation of public identities is not static and singular but dialectical, finding room for the many in the one, where what is universal is the encounter with differences both within and without.

LOVE BEGINS WITH DIFFERENCE

There is significant divergence of opinion between the ancients and the moderns on the experience of otherness in love. Plato presents the experience of otherness as a lack of something essentially needful, a kind of lost unity that is intimated only in transcendence, such that the experience of love, of being held out towards a perfect unity within which we belong, can be satisfied only when all particularity is abolished. This becomes a pattern of projection where particular loves are sublimated into more whole forms of transcendence, moving from the love of bodies, to love of country, to love of the cosmos, marking the steps on a ladder of increasing ontological surety moving from the necessities of economic life, to the realities informing political life, to the truths of philosophy and the life practice of learning how to die. In this ascent mapped out in Diotima's ladder, each lower rung and specific attachment to what is at that level suggests its own incompleteness and thus its own transcendence. What is desired overall is unity, though the experience of desire itself persists as a tension between the universal and the particular. Though it is all too easy to present a caricature of this ascent as a suppression of the erotic impulse, the movement toward transcendence is most definitely an erotic ascent for Plato. After all, the usually circumspect Socrates presents himself in the *Symposium* as an expert in erotics, registering the quality of eros as being animating by lack, longing, and the intimation that our being is held out beyond ourselves. And yet the prejudice against Plato inherited by an anti-erotic medieval tradition of interpretation and transmitted through modern thinkers wary of totalizing philosophies oriented toward a capital "T" Truth does bear some trace of truth. For if Diotima's ladder includes particular erotic fixations, attaching our loves to this particular person or community, those waypoints of the longing spirit are considered meaningful to the extent that the love of them transcends their particularities, seeing the one in the many and beauty in the beautiful, so as to move through that tension between the intimation of unity and the fragmented experiences through which it comes to presence in our lives.

To take a contrasting example from a modern thinker who is in some ways kindred to Plato, lighting upon love as basis for social bonds and personal growth, Irigaray defines our humanity not in terms of some unified form into which all particularities are absolved but as an essentially divided being, specifically divided into the two sexes. For this reason, she argues that human nature is two, and hence also that rights should be distinguished between those belonging to women and those belonging to men, as well as ideas that are inherently feminine and those that are inherently masculine. Her writing is a search for the former; in terms of masculine rights and masculine ideas according to her we have only to look to the canon, including the philosophic idea of a singular good and the political principle of the sovereignty of a people considered as a unified whole. For Irigaray rather, difference is inscribed into human nature itself. This pertains to the individual person as well, who exists as a person in relation to their own otherness. In this mode of comportment we are our own other. In temporal terms, we can become other than what we are, even in the absence of an end that would guide this unfolding of human potential, a process that is measured only by human freedom. Love is an experience of this freedom so as to have the world light up with the potential of change, guided by fulfillment of divided identities through identification with otherness, in a synthesis made possible inasmuch as one is held out beyond singularly given frames and identities, so as not to be pinned down lifeless by the frame of I=I.

The inscription of difference into the very biology of the human condition is what sets Irigaray's thinking on love, freedom, and negation apart from one of her key sources in Hegel. Her notion of negation is unique, in that it is taken to imply not so much a violent displacement of the other as a recognition of the limits of the self. For her negation puts us up against a world that is not of us; instead, while we are of the world, having been begotten of worldly mothers rather than of some sky goddesses depositing cosmogonic eggs onto the unbroken face of a primordial ocean; and while we are of the world, the world does not owe its being to us. The relation between human being and world is not perfectly reciprocal but rather one sided. So we owe quite a lot to otherness understood in the sense of the natural world, it having brought us into being, hopefully continuing to sustain us in our being, and outlining what will be the end of our being, with the life of the world outliving any one of us. This uneven relation, consciously grasped in the acknowledgement of indebtedness (say to one's parents), of precariousness (our fragile bonds with nature), and of finality (the earth will consume us all) is what makes us the dependent creatures we are; this disparity is also what makes us capable of love. And it is in this sense that the possibility of love arises out of the limits of the self, whereby the reality of otherness has the self stand back from the world that it owes its existence to.

This sense of distinction arising out of the genitive, with the person expressing themselves as a genealogy living beyond the dictates of either their historical past or given physiology, would be something like the first moment in Hegel's dialectic, that is simple negation, a free act that enters into the context of what is given while negating that in action, tuning the capacities of the world to new means and ends of freedom. The question for Irigaray, as for Hegel, is how to ground that distinctive purpose of modernity understood in terms of freedom within the lived culture of a people, whether by synthesizing freedom and *Volk* in an historical institution—for Hegel the state and laws—or reconciling freedom with the existence of difference, for Irigaray in the context of culturally mediated relations between the sexes. If Irigaray is less totalizing in the ambition of her project than Hegel, aiming at the ground of the human experience and its capacity to reproduce itself, rather than the sky level that Hegel chooses in describing the essence of those institutions that govern communal life—which are for him state, law, constitution, and civil religion[43] —then the gesture of intellectual reserve makes good sense, considering both the impact of any change to relations between sexes and the proximity of that essential encounter with difference in the day-to-day life of humankind. It is as if a change in that domain would be accomplishable, being contingent on the specific comportments of person to person and would immediately express profound influence, since such relations are the warp and woof of everyday social interactions, parlayed into the future by their distinctive capacity for regeneration.

In terms of the kind of progressive regeneration of a people that is the aim of both Hegel and Irigaray, our capacity for cumulative change, and thus for the historical dialectic of progress to unfold, is based upon a certain absence within every self. This is a kernel of negativity that is for both quite a positive quality of historical humanity, for upon it is predicated the possibility of stepping back from the immediately given, exercising this freedom from what is in order to enter into a mediated and differentiated relation with it once more. This ultimate aim of mediation though requires that the free subject be dislodged from apparent immediacies, whether of sense-certainty, tribal identity, or worldly suffering stoically endured. If the senses, the tribal collective, or a world of scarcity offer no trap doors for escape, if no alien god presents itself as a vehicle for transcendence, then it is that capacity of internal distancing or negation that allows the self to double back on itself, abstracting from circumstances so as to reconcile with the given on the grounds of freedom. In his preface to the *Phenomenology of Spirit*, Hegel describes this as a living substance "mediating with its own self its transitions from one state or position to the opposite," in a sense bringing the other into the self:

> As subject it is pure and simple negativity, and just on that account a process of splitting up what is simple and undifferentiated, a process of duplicating and setting factors into opposition, which in turn is the negation of this indifferent diversity and of the opposition of factors it entails. *True reality is merely this process of reinstating self-identity, of reflecting into its own self in and from its other, and is not an original and primal unity as such.* It is the process of its own becoming, the circle which presupposes its end as it purpose, and has its end for its beginning; it becomes concrete and actual only by being carried out, and by the end it involves.[44]

This pattern of the self-division of spirit promising mediation between self and other is what allows the transition of an "indifferent diversity" into a presence of otherness that is meaningful because it has become part of an identity. This occurs not through the mixing of a supposedly pre-established, original identity with some newcomer element, precisely because all identities are mediated, there being no "original and primal unity as such."[45] The mediation of difference occurs rather through acts of recognition, where a kind of identity is formed within which the elements that are reciprocally identified maintain their distinctiveness.

One witnesses this pattern quite literally when one recognizes another as a free being, not as being other as such but as a being that could become other than what it is, thus bearing out the regenerative capacity to change themselves. This kind of recognition would entail a deeper respect than the treatment of people as one would like to be treated oneself, because it would hinge upon an active expression of freedom, a gesture of unconditioned response, a dialectical opening operating beyond the logics of reciprocal stimulus and response.

At its most elemental and personal level, this is the kind of act expressed by looking another in the eyes. Recognition at the ontological level of persons begins with this opening gesture; indeed, in the absence of this confirmation of person to person relations, otherwise entranced by the movement of limbs and the portioning out of bodies in spatial relations, regardless of whatever niceties are exchanged in the tit for tat of conversation, it is easy to feel as if no real identification with another has occurred, however many stories of the day are traded back and forth.

For there is always something beyond reciprocation in an act of the recognition of recognition, in the *I see you seeing me* that unites persons as autonomous subjects in free relations with one another. Such a feeling of communion is possible because of that kernel core of being in relation to difference, where the subject marks the indeterminacy of an emptiness that can be filled in this way or that. In this sense, it is not the predication of certain qualities onto the subject that defines a person in their essential uniqueness but the absence of such objective qualities at the core of differentiation, with the source of difference consisting of a self that can become other than itself. This is what one recognizes in looking another in the eye, wherein an identification with the other is formed on the basis of their being different also.

Here one sees through to the other in the blackness of the iris, that portal into the psyche that subsists within a circle of predications, starting with the color of the eye and leading outward to all the conditional environs of the self. At the core of these of concentric circles with their layers of attributed objectivities is that question of what the other may be thinking, of what they may say or do. The recognition of another is a recognition of this essentially human capacity to become other, with a free being acknowledging another free being.

This freedom is that much more accentuated wherever two or more are gathered; that is, the political includes and amplifies differences. For the political is, in its essence, no singularity, because it exists in between people. The political is in this sense also, like love, always on the way; that is, it too is incomplete. Within this frame, as a groundwork emotion, love is closer to the essence of the political than fear is. Fear understood in the early modern sense is definite, serving as a benchmark for a society oriented toward security, with a fear of violent death composing the prevalent common sense.

If this is not a common sense for which the object of sensation is commonly apparent, beyond indirect encounters with the deaths of others, as a groundwork political emotion fear most certainly shapes the possibilities for association with others. In effect, if a relation with others is afforded, it is achieved through a sealing off of others who compromise the security of the self. Fear is localized in the fearful one and is in this sense singular (i.e., one can be alone and fearful).

The singularity of fear expresses a self-identical quality, in that there is a certain sameness to it, whether it is a fear of dogs or a fear of lightning or of heights. It is not a surprising, varied emotion in the way that love modulates between many different shades, with love of one's mother, love of one's country, and love of wisdom all being very different. Rather, fear corrals like toward like, while inclining toward strictly defined identities, with there being a monotonous sameness to fear and its effects.

It is precisely for these reasons that the early moderns from Machiavelli onward elevate fear while denigrating love, that is, in order to depoliticize the political. The aim of this substitution of fear for love as political animus is to bring a definiteness and manageability to public life by banishing love (and the fervor it can provoke) to the private sphere. In so doing the public realm is effectively eviscerated, there being no real reason then why one would want to participate in public life, beyond vulgar aims of vainglory or of profit for a clique. Restoring the political in its full dimensions, with all of its connotations of plurality, open-endedness, and danger requires somehow resuscitating the public value of love that for the ancients was to draw us into association with others, loving the good in common, however various the forms in which love of the other manifests.

NOTES

1. Georges Bataille, *DOCUMENTS*, 7, 1929. Translated by Dominic Faccini in *Undercover Surrealism: Georges Bataille and DOCUMENTS*, Dawn Ades and Simon Baker (ed.) (Cambridge: MIT Press, 2006).
2. Plato, *Theaetetus*, Fowler translation (Cambridge: Harvard University Press, 2006) 179e-180b.
3. Heraclitus, fragment §44, my translation.
4. *Theaetetus*,180c.
5. *Theaetetus*, 180d.
6. *Theaetetus,* 191e.
7. *Theaetetus,* 179e.
8. Paul Virilio, *Open Sky*, Julie Rose translation (New York: Verso, 1997), 138.
9. In an etymological instance where origins are located in a kind of onomatopoeia and identity of outward form and substantial meaning, Eric Partridge identifies an effectual base of the English word in the ancient Greek *barbaros*, literally "the Unintelligibles, the Stammerers". *Origins: A Short Etymological Dictionary of Modern English*, (London: Routledge and Kegan Paul, 1958), 39.
10. The now notorious recorded and leaked exchange between American forces killing Iraqi civilians bears out this crude form of logic. Dialogue between occupying ground forces and a gunship pilot speeds and simplifies acts of analogue judgment, worked out into a kind of digital language, one coded by a grammar reducible to a yes or a no, where discernment of "good" and "bad," civilian and target are blurred by a pixelated form of representation and behavioralist method of identification that could confuse a camera for a weapon. What saving grace remains

in the promise of higher resolution targeting would make the killing that much more automatic, and human exchanges that much more incidental, in a pattern of thinking even more robotic than this particular logic of identification:

> "There's one, yeah."
> "Oh yeah."

http://www.collateralmurder.com/en/transcript.html

11. Levinas, *On Escape*, 51

12. Slavoj Zizek on love as cosmic imbalance: http://www.youtube.com/watch?v=TJPhA9TGRls&feature=related.

13. *Genesis*, 1.2.

14. *Kalevala*, canto1.

15. *Beyond Good and Evil*, §172.

16. *Beyond Good and Evil*, §175.

17. A joke taken seriously, *as per* Friedrich Nietzsche, *Untimely Meditations*, R.S. Hollingdale (trans.) (Cambridge: Cambridge University Press, 1997), p. 79.

18. Levinas, *On Escape*, 73.

19. A delicate margin of politeness was drawn and an attempt at intercultural communication made in Canada's Arctic when Governor General Michelle Jean partook in a traditional Inuit delicacy of raw seal heart, seated on the floor together with her hosts, with an ulu knife for utensil and cardboard for a seat. http://www.cbc.ca/news/canada/story/2009/05/27/jean-hunt.html

20. I offer an extended critique of the Hegelian "end of history" thesis and the notion of humanity as negation that sustains it in *After the Last Man: Excurses to the Limits of the Technological System*.

21. G.W.F. Hegel, *Reason in History*, 56.

22. William Shakespeare, *A Midsummer Night's Dream* (New Haven: Yale University Press, 1965), II.i.64-73.

23. David C. Hendrickson, eulogy for his daughter, Whitney Demoraes Hendrickson, Shove Chapel, Colorado College, March 20, 2009.

24. Machiavelli, *The Prince*, Robert M. Adams translation (New York: Norton, 1992) 72

25. "It is imagined because the members of even the smallest nation will never know most of their fellow-members, meet them, or even hear of them, yet in the minds of each lives the image of their communion."

Benedict Anderson, *Imagined Communities: Reflections on the Origin and Spread of Nationalism*, (London: Verso, 2006), 6.

26. Carl Schmitt, *Theorie du partisan*, (Paris: Calmann-Lévy, 1972) 310; in Chantal Mouffe, *On the Political*, (London: Routledge, 2005), 80.

27. Following reflection on the horrors of millenarian cult-states and high-minded dictatorships in the twentieth century, this somewhat paranoid version of modern history would find resonance in the anti-ideological political theory of Eric Voegelin, for whom an overwhelming desire for the good as the cause of secular progress is somehow translated into the gnostic loss of faith in the world. [*Science, Politics and Gnosticism*, (Washington: Regnery, 1997) 58]. What is perhaps overlooked in such understandably nervous analysis (following as it does upon those tragedies of modernity) that would equate a guiding common sense of the social good with fascism or totalitarianism is that a concern with the common good draws limits around human action, introducing caution and respect into our doings with others within an understanding of our apparent differences.

28. "To found a great empire for the sole purpose of raising up a people of customers may at first sight appear a project fit only for a nation of shopkeepers. It is, however, a project altogether unfit for a nation of shopkeepers; but extremely fit for a nation whose government is influenced by shopkeepers." Adam Smith, *The Wealth of Nations*, (New York: Modern Library, 1937), 579.

29. "We became aware of the existence of a right to have rights . . . and a right to belong to some kind of organized political community, only when millions of people emerged who had

lost and could not regain these rights because of the new global political situation. The trouble is that this calamity arose not from any lack of civilization, backwardness, or mere tyranny, but, on the contrary, that it could not be repaired, because there was no longer any "uncivilized" spot on earth, because whether we like it or not we have really started to live in One World. Only with a completely organized humanity could the loss of home and political status become identical with expulsion from humanity altogether." Hannah Arendt, *Origins of Totalitarianism* (New York: Harcourt, Brace & World, 1950), 297.

In response then to Cavafy's question of what will become of us without our barbarians, the end of the possibility of there being outliers to the dominant society would seem, paradoxically to liberate the full potentials of inhuman barbarism.

30. G.W.F. Hegel, *Phenomenology of Spirit*, (Baillie trans.), 605.

31. I am thinking here of Hans Jonas' critique of the gnostic dimensions of existential politics [*The Gnostic Religion*, (Boston: Beacon Press, 1991) 320-40]; Eric Voegelin's indictment against 'immanetizing the eschaton' in a worldly longing that ought to belong to religious desire [*The New Science of Politics*, (Chicago: University of Chicago Press, 1987) 120]; along with Leo Strauss' position that progressive ideology diminishes the ethical significance of ideas of the good as emblems of critical thinking, retooling them into political imperatives for a "world state" that have "made perpetual and universal tyranny a serious possibility." *Natural Right and History*, (Chicago: University of Chicago Press, 1953) 23.

32. Homer, *Odyssey*, 8.287-394

33. Mary Shelly Wollstencraft, *Frankenstein*, (Boston: Bedford, 2000), 58.

34. Jean-Jacques Rousseau, "Discourse on the Origin and Foundation of Inequality Among Men" in *The Social Contract and Discourse on the Origin of Inequality*, anonymous translation, (New York: Washington Square Press, 1967), 218.

35. Jean-Jacques Rousseau, *The Social Contract*, 1.7, 22.

36. Derrida, reluctantly, on love:http://www.youtube.com/watch?v=dj1BuNmhjAY

37. *Symposium*, 193d.

38. *Symp*, 192d (my translation).

39. *Symp*, 191d (my translation).

40. *Symp*, 192e.

41. Homer, *Odyssey*, 8.320-324

42. Irigaray can be criticized in this regard for presenting a naturalized version of human sexuality that normalizes heterosexual love and marginalizes other kinds. As salutary compliment to this current of biologism in pyscho-social science, a dramatist may add something to the reality of the encounter. Aristophanes adds a third gender in his genealogy of sexual desire, tracing heterosexuals to a transgendered species of circle people, homosexual males to a male-male species, and homosexual females to a female-female type.

43. Hegel, *Reason in History*, Robert S. Hartman (trans.) (New Jersey: Prentice Hall, 1997), III.3.

44. G.W.F. Hegel, *The Phenomenology of Mind*, Baillie (trans.) (New York: Harper, 1967), 80-1. The emphasis is mine.

45. Interestingly, Hegel sees an element of a kind of negation even in the behavior of animals, "For they do not stand stock still before things of sense as if they were things *per se* with being in themselves [i.e. as if the things of sensation were unmediated, and simply given beings]: they despair of this reality altogether, and in the complete assurance of the nothingness of things they fall-to without more ado and eat them up." *Phenomenology of Mind*, 159. That is, animals objectify the world by taking it in as something edible.

Chapter Three

Barbarism New and Old

THE NEW SAVAGERY

> The danger is that a global, universally interrelated civilization may produce barbarians from our own midst, by forcing millions of people into conditions which, despite all appearances, are the conditions of savages.[1]

It is perhaps easy to see how the dislocating effects of technology have produced savage conditions for masses of humanity, as in displaced peoples, populations crowded into concentration camps, the landless dispossessed, or those sequestered to reserves; but what kind of barbarism belongs to the "universally interrelated civilization" that Arendt describes? What is perhaps most remarkable about the effects of the technological systematization of things, human things included, is the absolutely novel capacity for humanity as a species to communicate over the surface of the globe upon which we find ourselves dispersed. There is almost a trade literature concerning this acceleration of communicative capacity under contemporary technological conditions, interestingly with Canadian scholars prominently situated as offering a reflective account of the phenomenon. There is this steady legacy running through the history of Canadian political thought, from Innis's insights into the powerful political and social shapes given by our means of communication, whether they are wind-powered ships or canoes, trains on rails cutting a straight line across a country arranged as a corridor for the extraction of resources, or satellites in orbit over the planet so as to make every place potentially a center. This later, distributive sense of new communications technologies—that they have the capacity to disperse power—is noted in the works of Marshall McLuhan, who remarked at the re-tribalizing effects of digital technologies. It is as if in the digitizing of information, the "stuff" of communications, the medium itself begins to shape the limits and possibilities of what is communicated. Clearly, regardless of the scope and penetration of new communications technologies, both identities and relations remain mediated. For it is easier to simply assume that one's own identity is an intact given, as if one were born of the rocks and soil of a homeland, or likewise that one shares an immediate and ineluctably given set of common traits with others of one's own type, if there are limited reports to judge from. Ignorance is a pregnant ground for the formation of universal categories of identification. Hence the tendency of an intensification and expansion of communications to fracture identities even as it calls more individuals into constitutive relations with one another.

What is being suggested here is that it is precisely those ruptures in assumed identities that offer the opportunities for the formation of a more inclusive identity, that is a consciously realized universal wherein what is shared is difference itself, with a transfer into one's own sense of self-identity of that difference that may hitherto have been quarantined in others. Thus a philosophic imperative is made of a civil greeting; so "Who are you?" reflects upon "Who am I?" in a pattern that should be familiar to the world-traveler and the contemplative person alike.

At the intersection of the political and the philosophical, the virtue of courage figures strongly in inquiries into otherness, because at every such juncture what is pulled away, fragment by fragment, is the surety that an identity, whether of another or one's own, is known as a given. So, what constancy is not represented in the "things themselves"—where those things themselves are "persons themselves" (a nonsensical notion in that where there are no others there is no person, no persona, no mask, and no hidden, essential self hinted at behind the mask)—in such inquiries into the human essence the source of identification is supplied by the spirit instead, by which is meant not some individuated and eternal soul substance but rather the driving life force of a being that moves it forward toward engagements with others. Courage is a call toward the danger of fragile interdependencies—towards the trust in a friend, or in any other freely chosen association—holding oneself out towards a fated world that at times seems careless with life, human and non-human alike. To recognize this perilous condition that life is always in is to adopt that same carelessness toward life that is witnessed in the world in the act of reciprocally putting one's own trust in others.[2]

This specific sense of courage as a trust in others can be witnessed with frightening clarity in the work of war reporters, those who seek to bring back stories from as close to death as possible, like Odysseus looking for a wisdom in Hades that would guide him back home,[3] through the world somehow "out beyond," stitched together by the wakeful dreaming that is perception—what is also the stuff of poets. Or to carry on with the comparison of war journalism and myth, there is the motivation to see all in one view something that is inherently fractured and chaotic, as when Homer slips into the first person and admits that we humans "hear only reports and know nothing"—the first epistemological impasse in the West's recorded history, only solved by the invocation of the goddess Muses, who are privileged because they are "present, and know all things."[4] Isn't this just what the embedded reporter is looking for also—that sense of pure presence, specifically of an event or person or people, as communicated in a story? Here the note of veracity is cued by the reporter's being a part of the event itself, much as the Homeric pantheon was considered real because those gods were involved in the confusion of worldly life (e.g., they were fighters alongside of men, as well as taking on the form of humans and regenerating with them and enjoying their sacrifices). Thus one knows that one approaches the reality of things, and especially war, to the extent that the noise and clamor deafen, the smoke envelops, and human sensorium is indentured to a chaotic, unknowable world. If instead, the event seemed like a calmly ordered chessboard, as if seen from above, it is then that it would seem like fiction. What makes our war myths real rather is their direct contact with a profoundly irrational element, the jostle of a world of violence, translated directly into the camera's shake or the percussion of a frightened poet's meter. This, after all, is why the war reporter needs to be courageous, because their being is ultimately at stake in their capacity to tell the story that they took part in.[5]

Given the raw power of such stories from the front, it is not surprising that the first-person singular voice of the "I have seen" has not been co-opted for other, even more explicitly mythic designs, so as to attach that sense of the reality of war, experienced in direct contact with violence, to larger narratives, framed as cosmogonic conflicts between order versus chaos. Thus, the confusion and world-destroying qualities of the event itself are apparently reconciled with an order that somehow stands back from the fight so as to give it direction and historical purpose. So the reporter, with her beating heart, tear-clouded eyes, and ringing ears is embedded within an institutional narrative that would give her story meaning and context; she is rescued by one set of forces that then functions as savior, while the enemy is consigned to the darkness of a warfront at once distant and immediate to the ethical imagination.

And yet, isn't the refrain to this common mythic trope of a cosmogonic conflict between supposedly distinct forces of order and of chaos that there is, in reality, no distance between here and there? The logic behind this sympathetic intuition, grounded in the very technological phenomena that it would critique, is that each of us is embedded, each in our own ways, in a whole matrix of interdependencies that make that chaotic otherness that is the core of conflict at once a part of ourselves, such that the distant war does not function as a product of pure chaos so much as it is an unintended effect of global patterns of integration and ordering, now no longer so evidently efficient or self-sustaining.

Reflecting on the recent shift in modes of violence, from inter-state conflicts to violence within communities—especially the seemingly episodic outrages of ultra-violence by individuals—what is clear amid the chaos of those dramatic acts of carnage is a radical disconnect from reality for the agents of violence. And yet at the same time, that disconnect is systemic, such that copy-cat violence and terrorist franchises disperse what are otherwise marginal views: that a culture of nihilism justifies murder, or that personal fantasies trump the barest recognition of the existence of others. What is clearly absent in such extreme, yet increasingly persistent examples is any real recognition of others, who instead become the dramatic extras in another's act of gnostic violence. What this undermines socially is the sense of trust in others, and in social reality as such, with the new reality remade as a surreal set of reversals, with the expectation of violence becoming the new normal, global war on-going, torture normalized, outbreaks of zombie-like ultra-violence considered more and more the usual state of affairs.

AN ONTOLOGY OF DIFFERENCE

Ontologically the basis of reality is like this—a set of signifier-like perceptions in flux, grounded by a nature conceived as indifferent resources, a neutral field of ultimately chaotic relations guided only by a species' self-overcoming will to power, now brought to a moment of existential clarity in a humanity that is bound to nature precisely by our own self-made excesses. For isn't nature so much like this to us also, a spilling over of life into an historical manifold of forms, an evolving, increasing plurality, which has its limits in its environments, both self-made and cosmological, and finds the possibility for co-determined evolution in whatever niche the earth offers up energy for growth?[6] Species evolve in undetermined ways, by mutation, in response to randomly changing environments (think of the history of species, punctuated and spurred by asteroids and ice ages—life circumscribed by the necessities of physical reality). In this sense, what order one does witness in the seemingly perfect, interrelated forms of species is made possible by the random genetic mutations that brought forth that diversity of forms and the episodic, cosmic catastrophes that alter the environment at large. What is unique and truly significant to the present is that one particular species can have a hand in these catastrophes and these mutations; and further, that the apparent stasis of its particular form and its purpose, what we might call after Aristotle the *phusis* of the human species, is directly linked to the underlying chaotic plasticity out of which all forms arose and, in the most immediate sense, in an historical instant of crisis, with the outward, formal purposes of the species giving rise to conditions that will alter its form and future purposes, along with the potential viable forms of other species. It is as if historical evolution itself comes to willing consciousness at the moment when a species wills the crises that will occasion its own end. Or in other words, it is as if nature subverts its own determinations of chance and necessity by making these the products of mindful willing and so then needs to adapt to the chaotic after effects of the progressive rationalization of environments taken to its radical conclusion. Thus, in historical humanity one bears witness to the extreme excess of action in nature, that nature would produce a creature that tests to their limits the capacities for natural regeneration.

Is the human species in this sense nature's other? It is an attractive idea, according to a familiar trope, to simply invert the assignment of otherness from nature to man instead. So then it would be humankind that is somehow on the outside, with nature switched from being the externality—simply stuff, or objective reality, which is somehow thrown before man—to a being-in-itself and pure presence that has its own intrinsic values, independent of what the human type may antithetically impose upon it. Perhaps, according to this somewhat vindictive ontology of nature, there is some hope that by reversing the poles of alterity, nature will somehow come out intact (although indifferent to the state of humankind) while human beings will be taught a lesson in techno-ethics by the smug constancy of a self-correcting, harmonious system, as in *look what you can't do—you can't make me of you.* What saves this terrifying critique of a species capable of self-immolation on a genetic scale is that it too can be reversed, and in the positive sense, as in *look though what you can do—you can make you of me.* The first attribution of identity is impossible because it requires the perfect consciousness of intent in order to make a *verum factum* of oneself; that is, a truth known by identity with the self extended into the world through its artifacts can know only to the extent that it knows its own purposes fully and completely. Since this is never the case (for example, humans have their histories and will be part of the histories of others also; likewise, each decision is informed by a matrix of subconscious drives and half-remembered genetic inclinations that are consciously known only when they are manifest, and even then mutations are always possible), the second option for self-knowledge—knowing the "you" in the "me"—is more likely. Comparatively here we have, if not a methodology, then at least a question to ask on the way to self-knowledge, with the "who am I?" fragmented into the constitutive relations we have with others. It would then become extremely important both for the integrity of the self as well as for relations with others that we choose very carefully and place great emphasis on our trusts in others.

KNOWLEDGE BY PROXY

In the lead-up to the preemptive (and in the end, preempted) genocide of the island of Lesbos by Athens, in defense of their assertions of sovereignty and secession from the Athenian empire, envoys from the city of Mytilene say that:

> . . . we know that there can never be any solid friendship between individuals, or union between communities that is worth the name, unless the parties be persuaded of each other's honesty, and be generally congenial the one to the other; since from difference in feeling springs also difference in conduct.[7]

At the other end of history so far, we know from the benefit of hindsight that the Mytilenian position expressed here is more than a little naïve, in the sense that in this specific instance at least it was precisely the likenesses between the Athenians and the Mytilenians that brought about their war with each other. At the root of their conflict, it was their shared, preeminent desire for sovereignty that led both to anticipate war. Here there is latitude for difference in interpretation from the text, in that the Mytlenians begin their preparations for war after witnessing the slide from Athens's more or less benign, or at least accepted, hegemony into an explicitly imperial project, whereas the Athenians launch a preemptive defense of their democratic empire upon the warning of paid proxies, or *proxenoi*, that their power was beginning to fray at its edges.[8] But in any case, the two expressed desires are for much the same thing, that is sovereignty from any outside influence, a drive which is radicalized in genocide.

The question of difference arises when one considers just how that sense of "the outside" that is to be abolished might be construed and constructed. In the case of relations between Athens and Mytilene, is the margin of inside/outside traced out at the limits of Athens's ostensibly democratic *imperium*? Or does that limit have its concrete instantiation at some geographic border, say the walls of Athens or the shores of the island Lesbos, each a penultimate terminus and limit of the respective city-states? The last division may be the most obvious answer, though reductionist, since it is by no means settled that separation would imply enmity, considering that some of the island cities, namely Tenedos and Methymna, remained allies with Athens throughout the revolt. And notably there were elements within Mytilene that sided with Athens also. The walls that Mytilene put up to define the extent of its sovereignty could of course be and were in fact pulled down, there having been nothing natural about them as such. Rather, what limits there are to sovereignty have fundamentally technical grounds. If politics, conceived in terms of a free association of autonomous peoples distinguished by their differences is to have some properly ethical grounds, these cannot be reduced to this crude drive to power, though that be supposedly shared by all. We know what happens when this sort of reduction of political agency to crude drive occurs: namely, technical potentials for domination replace the negotiated acknowledgment of the terms and conditions of plurality as the defining source of political boundaries. It is as if the driving insistence to realize technical capacities occasions a certain political corollary in a reflex assertion of sovereignty to its limits, though in a way that would undo the fragile basis of the political as such, with constitutive differences obviated by a universal desire for power understood as sovereignty.

The really dangerous desires are in this sense the ones that are known definitively and that can be so known because they belong to the deepest substructures of subconscious thinking. It can hardly even be called a desire, if one knows its object so well that the desire is automatic, more a drive than like love, and if all share in it universally. What saves desire properly so-called is its inscrutability, that one is held out toward something at once essentially needful and pertaining to the other, for it is then that the autonomy of the self or of a people can assert itself in a free relation to difference, which must include the capacity to become other than one already is.

Freedom in this transformative, political sense can be discretely and provisionally understood as a freedom to act; at least this much can be known definitely—what is freedom fully and properly so called would have to be imagined anew every time that kind of fulsome freedom becomes possible, with each and every time being precarious by definition.

It is in this spirit of liberation to—understood as the desire for autonomy—that the Mytilenians give their reasons for originally allying themselves with the Athenians in a federation of free Greek cities contra an invading Persian empire, defending against "the barbarian,"[9] saying:

> . . . we did not become allies of the Athenians for the subjugation of the Hellenes, but allies of the Hellenes for their liberation from the Mede[10]

Who were the Hellenes then, this formative grouping of people as civilization, here described as civilized in a shared opposition to subjugation? Outwardly, they were many. A manifold plurality of islands and walled cities, populated by settler farmers and port-dwelling merchants, pirate raiding parties and priests walking the beaches of River Oceanos. In a way the Athenians did represent what was universal about Hellenic culture, testing their specific claim to universality as a civilization when they fled their own city walls, giving up that basic hallmark of identity by distinction, wrapped up in the sense of place, which is by definition the kind of identity that functions by exclusion, and hence cannot lay claim to universality. What the Athenians did when they fulfilled a prophecy to quit the stone walls of their city for the wooden walls of ships was to assert that their identity was robust enough at the level of understanding their common animus to hold them together as a people, even in the absence of a singular place from which to locate and pin down that identity, delimited as if it were some kind of geographic shell of identity.[11] There is something especially courageous about a people confident in who they are even in exile; for the Athenians, this would mean cueing into their capacity to adapt as their defining feature. It is in this spirit of finding shared purpose in flux itself that Alcibiades invokes a kind of patriotism unlinked from traditional touchstones of identity:

> It is thus that empire has been won, both by us and by all others that have held it, by a constant readiness to support all, whether barbarians or Hellenes, that invite assistance; since if all were to keep quiet or to pick and choose whom they ought to assist, we should make but few new conquests, and should imperil those we have already won. Men do not rest content with parrying the attacks of a superior, but often strike the first blow to prevent the attack being made. And we cannot fix the exact point at which our empire shall stop; we have reached a position in which we must not be content with retaining but must scheme to extend it, for, if we cease to rule others, we are in danger of being ruled ourselves. Nor can you look at inaction from the same point of view as others, unless you are prepared to change your habits and make them like theirs.[12]

In commentary on modern forms of imperialism, George Grant observes that our particular technological civilization is unique in the extent and scope of its destruction of other civilizations. It seems to be a kind of radical fulfillment of the core imperial drive that is intimated by Alcibiades, where the security dilemma is hypertrophied, thought through to an end that would make universal genocide seem like a rational conclusion to geo-politics. This has in some ways been the end put forward by the autocrats and fascists of modern global politics, at times inspired by racist prejudice, at other times by ideology, but always cueing into an existential instability that would see the very existence of others as threat to one's own existence. After the possibility of singularity is opened up as technological potential, the political changes its contours fundamentally. The age of genocide may be traced back to long-standing compulsions, such as fear, honor, and self-interest; indeed there are certain tensions that subsist over time within our historical humanity, whether these core fragments of human nature subsist recurrently or consistently. And yet given the radically new potential of historical humanity to communicate and act on a global scale, a unique range of expression is given to the possibility of making others act as one likes, so as not to become like them in custom and in habit. This kind of closure to the experiences of difference and of interdependency arguably has an impoverishing effect on the kinds of peoples and communities that result, considered in terms of their differences. If some kind of imperial security is won through the erasure of those essential differences, each conceived in terms of its cultural-political salience, then it is a security best suited to the dead, with life conceived as a rather more precarious source of well-being, resource and peace, that is a fulsome, living form of security.

Those whose thought is formed within tribal cultures can see through the kind of reductionist security mandates offering protection from others, because any essential form of connection to others shows up a zero-sum logic of winning at another's expense as a peril of its own making, with the roots of the insecurity being in the cultural formations from which the human world is composed. Homer "does not even use the term barbarian, probably because the Hellenes had not yet been marked off from the rest of the world by one distinctive appellation."[13] The Acheans were but many tribes gathered at a beachhead to pirate and raid, the Trojans but many peoples gathered for the defense of their host city. Here, a supposed categorical understanding amongst a self-identifying group is grasped (by the people themselves—the difference itself would be meaningless to an 'outside' observer) via their concurrent opposition to another. Out of this all-too common mode of identification arises a basic paradox: that a truly universal identity is only possible in the absence of any singular identity. A lesson of omission by a tribal poet, where a sense of civilization remains in the absence of a concept of barbarism.

THE LIMITS OF IDEALISM

One can tell apart the potential relative trajectories of humankind—the divergence of the species from itself—as distinguished in kind, in terms of those who hinge their judgments on human nature and those who do so guided by a sense of historical humanity: nature and history, each representing in itself a potential moment of historical change, whether toward, in the first instance, the necessary, or in the other potential, toward some attributed good. These two moments diverge and relate in the historical unfolding of the human species, sometimes projecting beyond its given terms and conditions in a cacophony of forms, other times constrained by the concomitant need to define and rein in the exceptions. Thinking on an historical time scale (a mode which is itself a rather wild projection beyond our given sensoria) one can imagine a kind of grand liturgical cycle of release and restraint that alternately pumps the genetic store up with a wild profusion of potentials, a surplus of human material, only to cull the stuff out of which we are collectively made down to a more useful, hardy stock, if only to keep the species to itself and capable of regeneration; or in more humane terms, to maintain the capacity to love our own.

This sort of cycled version of progress, marrying speculative idealism and ruthless realism, has presented itself before in Nietzsche's critique of Hegel's history, and in the conditional value that he places on the idea of progress in what is really a far more radical historicism, de-linked from any immanent, world-historical telos inscribed in the order of reason itself. Nietzsche's histories alternately preserve, venerate, and destroy, depending on what life needs and projecting toward what is beyond need. A similar pattern of history moved forward in pendulum swings—stepping back in order to make leaps forward—is presented in an altogether more obvious way in E.H. Carr's critical theory of progress, which has the progressive spirit drawing alternately upon the resources of idealism and realism.

Beyond the trivial instrumentalization of theory, that is, the conditional usefulness of the two perspectives, there is what we might call a dual ontology at play in Carr's version of history and its animus, with, on the one hand, reality consisting of the powers exerted by humans over humans, and how that power is perceived, and, on the other hand, reality consisting of the alternative realities that human beings can imagine and create for themselves. When it does work, idealism for Carr paradoxically issues forth in rather imperious attitudes toward the given—including given human nature—as incidental obstacles to be overcome. More often though, outside of revelatory historical junctures, idealist inspirations are simply steamrolled by power-seeking realists, with abstract institutions, like the League of Nations, simply falling apart in the face of self-interested actors. In a mirror image though of the lacuna in the idealist perspective, realism has a kind of inherent impotency, in that by simply describing the world, realism offers no incentive to change it, nor can it offer any convincing descriptive account of change. What an obsessive focus on some underlying reality can do though is to serve as a ground-clearing force, exposing the play of interests behind ideals, and getting rid of ineffective institutions. What we need according to Carr is to reestablish a tension between ideals and reality, so that our ideals have substance, and our reality, a sense of purpose.

OTHER THAN REASON

This kind of polarity between a realm of ideal freedom on the one hand and of necessity on the other is familiar in the history of ideas; the concept would have one thinking back to Plato and his distinction between the good and the necessary, or even further back to Parmenides and his distinction between persuasive truth and mortal opinion—between what is in and of itself, and being that is dependent in this way or that. The question and point of distinction within this perennial distinction pertains to what relation there is between those two poles. How do we read Parmenides's aphoristic proviso to his identification of truth and being, a qualification of that exclusive identity that would have us study the realm of phenomenal appearances via the mortal, changing opinion of human kind, in all its variances and moments of being at odds with itself? The impasse between truth and appearance arises because, as he writes:

> . . .the things that appear must genuinely be, being always, indeed, all things.

It is an attractive notion to think of the world simply and exclusively (at least in an epistemologically rigorous sense) as being a set of signs. We find this kind of totalizing semiotics in Barthes and in Baudrillard, where the world glows as a set of signifiers, though free floating and infinitely re-arrangeable according to the dictates of style and interpretation. For if what is available to be known is solely and exclusively through the appearances of things, then all of the world becomes a cipher for meaning, however prone to reevaluation that world may be. Thus, for Baudrillard, the sandstone spires and ocean-sculpted canyons of the Utah desert are themselves a kind of script for geological history, like the scat of time passing, that offer up a silent reflection that dwells simply in being and appearance and goes no further. The question of origins is, in Baudrillard's *America*, reduced to a question of process. Some slight revolutionary *mythos* may resonate, but it is deflected into an engineered process of self-transformation or of the management of things, as if technical engineering and the rhetoric of self-help would replace revolution. In this turn of historical consciousness, the world goes silent, all the while fully disclosed as information for the interpretive awareness, with little left for reason to do other than bear witness to the world's auto-cannibalism.

To counter this apparent historical tendency, one can ask what relation does reason have to its own other? What bearing does true being, somehow beyond appearing, have upon distinctively genuine appearances, that they would stand out from the manifold, even if truth itself cannot come to appearance short of being itself conditioned, to be a truth in this particular time and place—a truth for them or a truth for us?

Perhaps the first question cannot be answered or perhaps it already has been, though with the question itself forgotten as a stepping stone left behind. Perhaps this othering of the self is already endemic to a technologically advanced civilization—of whatever ideology, capitalist, fascist, or socialist, or so on—that is, perhaps it is no longer a question of enlightenment that propels the human species to change into its future but rather of an auto-correcting/auto-mutilating adjustment of the species to self—made environmental circumstance, such that what civilizational reason remains is brought to the level of a cultural artifact and trace of an already forgotten experience. What if the real cause of progress on a global, systemic scale is actually forgetting, as in the undoing of the Western destiny to progress, to be replaced by a knowledge that is uncaringly ignorant of causes, effects, and attributes? By such a turn, the basis of knowledge would be epistemologically reduced to an array of potential ciphers—that is to representation as information—or better, as informational indices, like the *metrics of conflict*, whether those "metrics" are either known or unknown. It really does not matter which, so long as there is an index for whatever is known or unknown. Could one measure nothing then? Has it been worth importing that number, the empty set, into language? What would it mean to forget zero, to do away with any sense of nothingness itself? The timelessness of technological development, its unfolding simply as a succession of nows, or just as blithely, as a catastrophe strewn landscape of apocalypses—in Heideggarian language, the un-concealment that is the end-of-time concealment of any possibility of authentic coming to presence—seems to obviate just what consciousness of absence provides: that is a notion of a beyond, and of love capable of grasping it, and the imaginative reason needed to intimate it. Think of the outrageous speed that is late modernity's hubris and compulsion.[14] Perhaps in order to get beyond that particular, highly imaginative, yet still determining code, modern civilization would have to make a myth of itself, that is to be forced to give an account of its origins when it has forgotten its own originality, thinking that an immediate rationality prevails, and hence that origins are no longer of significance. For within the context of a prevailing systemic rationality, the questions of origins and of originality can only represent the breaking off of significations, which have their basis in a matrix of immanent relations. At this point, all representation trails off into that which cannot be signified by something else, precisely because it is unprecedented. Where there is no lockstep relation to preceding events, in that state of unprecedented, unconditioned potential, reason has no given facts to bear down on, no world of objects to measure, and short of reflecting on the value of reason as such, the cosmogonic spectacular initiates a turn to myth and the undoing of established meanings.

For isn't the issue now not a general unthinkingness but rather too much thought, too many contrived systems embedded in every object that surrounds us, wired as we are, each at a connected center of a rationalized world? In this context, one can easily imagine a thoughtful zombie; a loving zombie would be more difficult to conjure.

In the truly caring desire to be mindful—as in the widespread Western interest in yoga, for example—what is actually experienced, paradoxically, is precisely the desire not to think, not to analyze every experience as simulacra, not to be existentially anxious in the presence of other beings, including oneself. (An archetype for this anxious mode of being would be the anti-hero and hitchhiker to the universe, Arthur Dent, who though being able to fly by telekinesis, can still manage to feel uncomfortable in the presence of a potted plant.) Thinking action through as process is now a post-Fordist form of recreation, something done by encultured habit, the social product of lives lived out of and into the monitor.

BEING IS INTERDEPENDENCY

It is manifestly apparent that our human world has become an engineered mess, brought forth and sustained by patterns of systemic destruction masquerading as rationality, such that what is needed is to unthink those apparent necessities of systems of integration, to unthink our barbarians as automatic others and to rethink ourselves in free relations to the differences we depend on, that is to make way for an ethic of difference by restoring thought to its origins and pathos in an absence of any singular identity, thus borne out of a need for love and of being toward others.

The experiences of love, as well as the hatred of others, have at their core a certain dependence on the other, and one that is reciprocal, in the sense that both kinds of identity formation—both the loving and the hating—bring the other into an identity, whether the identity is one of loving need for the other as essential to one's being, or if it is an identity of violent contrast, where the "we" and "they" identities are formed in opposition to one another. In each of these kinds of relationships, the relation consists in an other over and against whom an identity is held out toward in its formation: an other that one is not but who one needs. For of that other being—their being what one is not, in the senses of both their complementary qualities—as in the case of love—as well as in the violent displacement of identities—as in an-other that others, finding its relation to its own other by killing its other—in all of those instances otherness consists in the certain absence, an opening left for freedom, literally for being otherwise and becoming other. This essential opening consists both within and out of oneself, that is both in the capacity to change inwardly and in the capacity to freely represent one's own differences, as in an identity conceived in relations with others. For left entirely to oneself—unworlded as if ejected into the vacuum of space, with no spaceship nearby, not even the one that ejected the lonely cosmonaut—what cue would there be that the self is not a whole? It is that disjuncture within the self-satisfied aspect of consciousness (the later perhaps proper to an infant age of humankind, yet all too endemic to our present global techno-culture) that leads one into free relations with others. The freest relation that one can have to another is a loving relation with a freely chosen other. This would be the ideal over and against which, at the most conditioned opposite extreme, would stand the hatred of others, with the alternatives then in between an automatic other, imposed by their sheer presence of others, and a freely chosen other. Love of others and hatred of others both begin with that absence—the basic presence of the other in whatever sense, whether they are considered to be what one is not or what one is. Both modes of relation are forms of acknowledgment, proceeded by different emotions—directed toward either fulfillment or despair—of those basic conditions of human existence, that is considering difference and the fragile connections that subsist between self and other. To which one may respond, whatever barbarians we may see in some others, it is those barbarians that we become ourselves.

READING FOR PERSPECTIVE

The encounter with difference takes on a note of precision when one comes up against the thoughts of an ancient philosopher, who is themself not present to clarify or rebut a misrepresentation. Here one knows that there is something not fully understood, something that exists independently of whatever one may think of it, something that in the silence of an absent author's text keeps to itself with the self-assurance of a dead man's eternal reflection. The ever-so rigorous Straussian scholars aim to interpret these kinds of hallowed texts in such a way as to understand the ancients as the ancients understood themselves.[15] What is obvious on the surface of this method of interpretation is that the measure of self-knowledge held up by authorial intention can apply only to how the author would have understood themselves, not to how they would understand themselves presently. Part of what the thoughts of the ancients represent then, in the present tense, are limits that distinguish contemporary patterns of thought, revealing the currency of common sense as being particular in its scope and significance. In a genuine, open encounter with ancient thought it is as if our foreground knowledge is thrown into relief against a blackened backdrop, like the moon against the darkness of night, with what is known registered against the limits of knowledge.

The sense of modesty that comes out of an open encounter with a thinker from a distant time and place is akin to the respect accorded to a person or a people alien to one's own civilizational expectations. Even before the particulars of difference are introduced, what is registered is some particularity to one's own civilizational position, which may have, prior to the encounter, seemed so self-evident and universal as to be a given, underlying, and common-sense assumption. At this level of first impression, the other—whether they be author, or person, or people—invites critical reflection on the premises of identity. One thus begins to question identity and know one's limits, such that an abstract identity of I=I, assumed in relation to some independent third term, is replaced with identity borne out of a relation to difference.

At an extreme of otherness, beyond the absence of the author themselves there are those instances where even the text becomes silent, like a wilderness buried under concrete. In the purest cases of absence in the encounter with difference—and there are so many of these that there seems to be only one singular nihilation of apocalypse past—there is nothing left for reflection to take as its substance. In these encounters there remains an abyss of meaning left to fill in, such that when we imagine what is lost to historical humanity, we wonder at the depths of a cumulative imagination silenced by time, with the nothing that presents itself as having been here before provoking a sense of loss beyond salvage. The unremembered death—an historical genocide of forgetting that awaits every peoples and every person, living or dead—confronts the reflective imagination faced with fragments of the past, seeing in them the surety of one's eventual undoing also, and thus the fragile contingency of their being like ours.

Think, for example, of the passage of a canoe over a height of land crossing, what the Nishnawbe Aski First Nations who inhabit the Hudson Bay and James Bay watersheds call the "small places": mediating routes that connect the larger fur-trade rivers, with their still well-marked portages and disused Hudson Bay trading posts, in order to pass over heights of land from one river's watershed to another, thus securing traditional, nomadic iterations of travel back and forth from seasonal camps—a pattern of movement very much distinct from the linear traffic of goods into and out of the interior.[16] Whereas on the larger trade rivers we have fragments of a history—trading posts and ports—the little ways in between have in many cases been abandoned from use, such that the routes themselves have seemingly disappeared even after only two lost generations. Routes that may well have simply been the results of human traffic over what may have originally been animal trails grow back in when people stop following them. Where there is nothing built to aid traffic, nothing remains to indicate that there was a way there, so long as the way itself came to presence as an opening, a kind of originating absence. So, where one would have located a canoe portage way by a break in the trees rather than a constructed railway for the passage of York trade boats, all that remains is its absence, and at the extreme of nihilation, an absence of absence.

This perspective changes the status of historical reflection, from a gathering together of fragments, in order to synthesize some sort of probable whole, to a fixing upon certain lacuna, gaps, openings, as places of access for a creative historical revisioning. If all that subsists of a little way, say going upstream from one beaver pond to another, is a canoe-wide channel in the boulders, made that way by spring ice flow or the hands of earlier travelers, it is impossible to say, then coming into some sort of connection with that past and the now silent people behind it means moving through those passages oneself, putting oneself in the place of the absent other. This is no objective, synthesizing recreation of the past from its parts but rather a way through an apparent absence, with all of the connotations of incompleteness, longing, and foreshadowing of one's own fate that a silent history foretells. The modern canoe traveler looking to follow in the footsteps of peoples who haven't left tracks through these little ways for generations, wishes to wake up listening in the wilderness for the voices of the dead; and when he hears nothing, when he is lost, then he knows he is getting close, having by extraordinary effort pulled away from the gravity of the paved highway and a lockstep history of technological identification and ordering.

The pattern of radical historicism leaves open the possibility of the undoing of a synthesizing history, and in so doing reserves the potential to recapture that original, undetermined moment—an alternative history that is radical in the sense of returning to roots, such that "the past invades our future."[17] What is being put forward here, I would suggest, is not some determining fate that would connect the past to the future through some series of causes and effects, whether immanent, divine, or otherwise, but rather the interruption of those kinds of successions of events, such that the future is freed from the machinations and expectations of the present. What is left in the relief of history from the narrative of progress, wherein the past was linked to the present as a kind of anticipatory gesture, is the possibility of the past being repeated, not in a representational sense (as in a kind of historical reenactment) but rather in a return to the original moment and an unprecedented past. The potential of a radical turn subsists within a past gone silent, borne out of a self-assured modesty that can recognize when it is lost, that can see where there is something missing, and that is capable of finding ways through the gaps that open up as things fall apart.

But these are most definitely radical instances, being so close to the root of things, and thus so far removed from the contextualizing passage of history, with its web of implied and evident interdependencies, such that it becomes hard to imagine a discursive outcome to those kinds of reflections, something somewhere in between the extremes of prophecy and wordlessness.[18] More approachable would be those examples where even a few traces of the past remain, projecting beyond the unconditioned original.

Every act of interpretation has an element of wonder at its origins, as one is held out toward a recollection transmitted so tenuously by paper, papyrus, or pixels that its very presence requires a kind of conjuring from the past, a code sent back over silenced times to a source that can still somehow speak to the present. This is perhaps not an altogether new situation, but it will likely seem new each time an ancient author is turned to again, as the experience of reading Plato for the first time, for example, must seem shocking and inspiring to every student of philosophy who finds their way back to that particular origin of Western thought. Still though, in the case of the Platonic corpus a degree of integral wholeness remains—there is a kind of body of texts—and in that something like a cosmos in waiting—a whole—remains within the pages of the extant dialogues, whatever questions may remain about the authenticity of some or the extent of what may have been lost to collective memory. The passage of time is not always so kind to philosophy and to memory. The love of wisdom loses its object; and even the fact of forgetting may be forgotten.

And yet even there, in those cases of drastic and irretrievable losses one may wonder—and this may always come up—whether that is all, whether the bare transmission of facts through time, passed along like electronic relays of cause and effect are all that remains, with the bare present standing in as signal child of a pregnant past. For it remains possible that even the absences in historical memory contain within them some truths, with an intimate acquaintance with those nihilated pasts preserved precisely in their absence. Love, after all begins with absence, with a loving being held out toward some essentially needful other, perennially caught between laughter and tragedy, fulfillment and loss. Philosophy is predicated on a lack also, with knowledge sharpened by its own sense of limits. That is to say that there may be positive presences in absence, attachments to an othered past that the effects of distance, strangeness, and radical alterity serve to sharpen, such that even the numbness of a careless indifference may contain within it twinges of misgiving, if only as lost opportunities for hope.

These are the kinds of pasts that invade our futures, finding a presence in the withdrawal of the present into absence. For there is also a sense of the past as that which must be nihilated, (*must be* even if it will not be) transcended toward a future freer from violent necessity, political isolation, domination, and human suffering. One can credit ancient traditions with a host of these kind of mythic pre-histories, offering familiar successions of human historical life, apparently fated to violence by their isolation from one another.

NOTES

1. Hannah Arendt, *Origins of Totalitarianism*, 302.
2. This quality of precarity coming through as an impact of war is given a salutary status in Judith Butler's *Frames of War: When is Life Grievable?*, (London: Verso, 2010), turning the security dilemma on its head as source of an empathetic social bond.
3. *Odyssey*, 11.
4. *Iliad*, 2.523.
5. Susan Sontag registers this basic truth that the veracity of war reporting must somehow express this mediating quality wherein the suffering of others has an impact on the one watching; see *Regarding the Pain of Others*, (New York: Picador, 2003).
6. Nature has an environment too! "Environment" is of course an infinitely repeatable, perfectly universal concept, as applicable to the enframing context of the natural world as it is to the human world.
7. *History of the Peloponnesian War,* 3.10.
8. It was a local *proxeni* who apparently informed Athens that a revolt was brewing in Lesbos. [*History of the Peloponnesian War*, 3.2] A *pro-xenos* would be a public guest—literally a sent out foreigner—a citizen in one land serving the interests of another place and

people. Proxies would have entertained ambassadors and advocated for a foreign city, enjoying both the privileges offered by their allied city and the protection of their own city's laws, remaining citizens of the city in which they lived, while bearing allegiance to another (C. F. Smith). This pattern of patronage would often be inherited, passing from father to son, (Liddell and Scott), suggesting a certain dislocation of identities, fissured from soil and shifting to blood ties and a prosthetic form of association, specifically here as a mode of imperialism.

Then as now, empires and classes within empires exerted their influences across a power gradient of various means, from invasion, genocide and colonization to the currying of favors with friendly locals, with these "soft" and "hard" forms of imperial politics shading into one another—as in this instance where a *proxeni* in Mitylene alerts Athens to an impending revolt, thus occasioning a deliberation on genocide, in this instance of the Mitylenian people by those who saw themselves as shock troops and bearers of Hellenic civilization. In that precarious moment upon the brink of an extermination, upset perhaps by an ethical conscience, the Athenians half-willingly decided to spare the people *en masse*, though with "rather more than a thousand" Mitylenians reportedly killed to quash the insurgency. [*History* . . . , (Crawley trans.) 3.50]

And yet, though sometimes serving imperial purposes and instigating violence, we have instances of these citizens of mixed allegiance working to defuse cultural tensions also. Though Thucydides identifies himself as an Athenian straightaway in his *History* . . . , the allegiance owed to his native city is definitely a qualified one, given that he was himself a *proxeni* to Athens. We know very little about the man himself beyond rare self-identifications in the text, there being no hidden essence of nationality behind the person to uncover; but what does come forward is a deep care for humanity that cannot be grasped in the formal, reciprocal terms of citizenship. One cares about the outcome of human affairs most acutely when the potential outcomes are not calculable in the terms of this for that, liberty for just action—that is beyond the bounds of lawful, regulated behaviors. Responsibility beyond the performance of simple duty is attenuated precisely when it becomes a political responsibility, and this implies a measure of the unpredictable, because others are involved. So in Thucydides' case, he would find himself caught up in a class war in his native Athens, though inclined to neglect the party divisions in Athens for a more reflective, conciliatory approach, one that could only come with distance. It is this critical distance that belongs to the recognition of difference as constitutive of political life, and leads to the acknowledgement that we are all, every one of us, divided beings and our own others.

9. Thucydides, *History of the Peloponnesian War*, 1.75.

10. *History of the Peloponnesian War*, 3.10.

11. *History of the Peloponnesian War*, 1.18.

12. *History of the Peloponnesian War*, 6.18.

13. *History of the Peloponnesian War*, 1.3 Thucydides compares Hellenic and barbarian customs at 1.5-6, distinguishing the barbaric custom of bearing arms as personal protection.

14. "The reduction of distances has become a strategic reality bearing incalculable economic and political consequences, since it corresponds to the negation of space. . . . Territory has lost is significance in favour of the projectile. In fact, the strategic value of the non-place of speed has definitively supplanted that of place, and the question of possession of time has revived that of territorial appropriation." [Paul Virilio, *Speed and Politics: An Essay on Dromology*, 133] Lest one respond by quite rightly pointing out the renewed importance of occupation, whether as a strategy for imperium or resistance, it is worth considering Virilio's opening question in the same essay, "Can asphalt be a political territory?" 4.

15. See Leo Strauss, "Political Philosophy and History" in *What is Political Philosophy?* (Chicago: University of Chicago Press, 1959), especially 76.

16. Jonathan Berger and Thomas Terry, *Canoe Atlas of the Little North*, (Boston Mill Press: Boston, 2007), 10.

17. Martin Heidegger, *Pathmarks*, (Cambridge: Cambridge University Press, 1998), preface.

18. Hans Jonas, *The Gnostic Religion* (Boston: Beacon Press, 1958), pp. 320-344.

Chapter Four

Barbarism and Civilization

ON OUR BARBARISM TOWARD NATURE

A natural human attachment to that most imposing and perilously present version of the Other—that is Nature—is shown in the two human natures: one nature desiring love, the other desiring reality. That is, it is natural for us to want to experience a kind of spontaneous beneficence in nature, as well as to see nature for its sheer reality. The reconciliation between the two modes of relation, the circle that makes human nature whole, is the recognition of our being held out toward an essentially needful other. In this sense it is natural for us to see faces in the clouds, to read nature in human form, not so much out of a projection of certain arbitrary assignations of humanity onto nature (that is, with both nature and the human species assumed as some kind of fixed archetypes) but rather out of a desire for an easy affinity with the sky, sun, and clouds. Similarly, it is pleasant to stare at the sky, to simply look at it, not because it is of human design, say as its pleasant to look on a piece of art or fine craft, but because it is simply present to be seen. Nature stands out from the stream of becoming as an ordering of its own making, though a making beyond making, an ordering that an eternity of nature-made storms could not wash away. What confuses this nature of things now is that it is not so evident that there is such a clear distinction between what is by nature and what is made by artifice, where the weather has origins in the unintended effects of human technology. Out of this confusion comes a strange predicament where humanity seems victim to its own powers, and where nature seems forgotten even as it becomes systematically ordered.

For nature as such cannot know nihilation the way that humans can, such that even the human sciences of nature have no language for a pure *nihil* but that of some assigned singularity, no end save in perfect unity, no undoing that does not promise some new start to things coming into being again. It would seem that the original division between the human species and nature as a whole holds open that relationship at least, with some space for future beginnings remaining at the fulcrum of human action into nature.

IS NATURE NORMAL?

Underlying the rising up of human action as abhorrence, as in the barbaric act or some other gesture of inhumanity, is the still prevalent sense that nature subsists, being somehow prior to the exception, being somehow more basic, more original than the things made by human beings. Alain Badiou puts forward an ontology derived from Heidegger's interpretation of Aristotle's *phusis*,[1] yet in an even more spare form of a mathematized set theory.[2] The starkness of Badiou's depiction of nature as multiplicity, where those multiples are natural to the extent that they can be considered normal, resolves nature into that simple possibility of there being a manifold. Nature is thus seen as the set within which a being can come forward as simply being, in very much the same the sense of Heidegger's translation of the *phainomenon* of *phusis* as a "shining forth." But what is the basis of possibility of a being coming into appearance as a being? For Badiou it is its normalcy as a natural being that lets it come to be so; I wonder whether this position can account for both the natural awe that arises out of natural phenomena as givens, as well as for the amazing plasticity of human nature, both of which present themselves as natures in so many different forms. One would have to somehow recognize nature as phenomenal, that is, as standing out from the norm, which is now the systematic technological, the new human-made environment. From within this, nature subsists as a standing forth in the closure of being, with nature persisting as that post-system exception beyond system, the code that would unlock its own evolution, a being with its cause for change within itself, and thus as exceptional in its genitive essence, with what is natural defining its own exceptionality, at once subsisting yet being beyond itself. For after all, nature can change; it is the self-representing phenomenon that puts itself beyond itself at every generation and within every event, with the possibility of difference being the ground of all natural unfoldings. Heraclitus says something similar: "The sun is new again every day." It is in this sense that nature can change without becoming somehow not itself, as in a digital thesis/antithesis relation, but rather by becoming other than itself instead, because it contains its differences within.

Such a position would be a departure in form and substance from what is, within late modernity, an Hegelian-derived conception of nature as thesis, that is, as a static, law-bound given that only the creativity of human technological savvy can break through. In Hegel's telling, it is the sheer dull givenness of natural being, stuck outside of history in its inanimate animate existence, that sets apart nature as given. "The universal, here regarded, is, only in the first instance, what remains identical with itself; its movement is merely the uniform recurrence of the same operation."[3] That is, to unreflective consciousness, nature is not yet grasped as universal to the extent that it remains the mere objects of sense experience, as mere sameness of the other taken in as "stuff," and of a sort of stuff that moves in recurrent operations, something like a primitive machine made of nature. Or again, in a more generalized presentation in the introduction to his lectures on the philosophy of history, it is historical man who "constitutes the antithesis to the natural world; he is the being that lifts itself up to the second world,"[4] that is, a realm of freedom that transcends the determinations of a nature somehow stuck outside of the spiritual flow of meaningful change, that is, change that negates and advances synthetically, shaping the substance of nature into forms that serve worldly human freedom.

It is here, at the origins of the idea of secular progress in German Idealism that the diminutive is attached to nature as the other to historical progress, as in an intensified disavowal of the early modern derogation of nature, i.e., in some supposedly "mere" nature, that is, a nature "solitary, poore, nasty, brutish and shorte,"[5] the backward life of humanity outside of the technological, and all else that is derogatory to the natural world and its simple, sustaining, self-moving quality of being.

What comes upon this flux of self-generating being is a singularity threatening all that is multiple and varied, whereby in systematizing what is those beings are coopted in their distinctions, with those differences becoming sources of interference and inefficiency in an emergent form of systemic auto-rejections—economic, social, and political—accompanying the rise of system management as social organization and the cult of efficiency as common good.[6] This is the systemic form of what in Hegelian terms would be called a "bad infinity," where the patterns of technological integration are compelled to unfold in a uniquely self-referential way, with those systemic auto-rejections providing the animus for historical change, and with the potentials for progress unredeemed. What we find instead is a corporate variety of system management inscribed upon the new terms of social organization—a new kind of tepid fascism working to oust democratic governance for a new socio-economic discipline, all the while rendering nature into resource for technological imperium.

The modes of resistance generated out of the natural manifold consist in terms precisely opposite to Badiou's "normal" nature, that is, the phenomenal and standing out form of nature. For though nature certainly does subsist as the simple remaining present of what is, it certainly does also stand out, where the presence of the natural opens up revelatory spaces within what is otherwise just environment, and where, within that environment natural and social determinants of being are themselves co-determinant. That is, nature comes to presence as being natural to the extent that there is, as Heidegger would say, a "standing-forth" of Being as *phainomenon*, etymologically speaking as a "shining forth" of what is beyond its simply being there. Nature has this astounding, awesome character to it; for if the technological ordering of things is essentially lifeless—a virus-like system imposed onto living beings, with efficiency describing the new alternative to justice—then the management of things is revealed for its crystalline fragility the moment that which is truly substantial manifests itself in such a way as a standing forth. This is seen in the spectacle of nature and especially as in the spectacle of disaster. And this is how the self-presentation of nature is in turn represented, as a turning back of substance onto the technological ordering of things; or as George Grant asked some decades ago, "Is the non-human simply stuff at our disposal, or will it begin to make its appearance to us as an order the purposes of which somehow resist our malleablizings? Are there already signs of revolt in nature?"[7]

This is the source of a nature that revolts within technological ordering, within what is a uniquely reciprocal relationship between technology and nature. That is, when nature revolts and the designs of technological dominance come apart at the seams, a kind of global spectacular results, as if the globe in its entirety in that moment becomes in some sense a conscious being, capable of generating reflection on and of itself. It is in this way, via the technological, that the natural can come to an at once integrated existence and at the same time produce individualizing action and distinctions within that global pattern of integration. Attentions are transfixed, days of collective focus glow with meaning and shared sense, and the world becomes as if it were entirely public, totally transparent; it is as if the world were to belong to one global public, in an instance of an imperium of sense that has lost its empire, and yet still remains as the new common sense. This would be a global solidarity arising out of what can be described as the new technological ecology.

Critically speaking and in ethical terms, this would mean rethinking our own barbarism first, in order to find some relation to what has been called the other, what is conceived most primordially within humanity's relationship with nature. For this is a most obvious form of barbarity—that is, human comportments toward nature within technological ordering, where unconscious numbness to what is becomes the grease on the wheels of an anti-human, anti-natural form of historical development, moving lockstep forward toward a future where corporate inanity in social organization seems to be the norm, and nature is treated essentially as resource. This is the horror and the danger of the technological age; it is the awe that we have made for ourselves as historical humanity through the action of technology into nature, a new danger requiring the delicate operation of a reverse engineering of nature back into technology, so as to remove the kill switches that humanity has now built into the planet.

For it is difficult to imagine a more barbarous act for humans to accomplish than to devise the means to destroy humanity and the planet *en masse*. I find myself agreeing with a commentator on the thesis that the barbarism of violence is more attenuated when it is on a small scale, namely prior to the innovation of the modern state, to which he counters that the absolute measures of violence cannot be obviated by per capita deaths per thousand, as one should somehow be pleased by one's individual positioning within a matrix of percentage chances of personal death.[8] The concern for peaceful relations with others quite naturally goes beyond those within the immediate enclave; after all, love is always possible, such that even the existence of war cannot undo the possibility of identifying with others. Inter-cultural relations understood in this sense begin precisely with a critical self-recognition of one's own situation and its limits, presuming that these limits in some way, whether conscious or otherwise, were what drew the party into the relationship in the first place. A free relation to difference thus begins with acknowledgement of the ends and limits of the self, finding the other at the source of the self.

Beyond this, the occlusion of the other takes over in a pattern of suppressed subversion, when self takes it upon self to isolate and redact any sense of otherness or of difference within, perhaps assuming some blood and soil identity, or reductionist history, anything that occludes otherness, mapping some supposed singularity of identity of I=I onto all the diversity found within any one culture or one specific person. This is when barbarism comes into being, as a projection of that otherness onto others, whereby "they" are seen as free of civilization to the extent that "we" are properly programmed by it, fed by the dull drone of filler programming as our inputs and human resources as outputs.

Notions of barbarism are thus sustained by these reciprocal relations between self and self and self and other. This is the symptom that is also the route to the cure, namely recognizing that barbarism is a relational concept, though one with an absolute measure in terms of its potentially genocidal consequences.

For what sustains barbarism as a particularly violent set of perceptions is primarily a lack of critical reflection, which even in small doses works to undo the bonds of ignorant self-importance, replacing it instead with a peaceful conscience. This would be the aim of inter-cultural communication, with that common core of respect for difference within others and within oneself being the code and pre-condition for a preternaturally natural, that is outstanding, emergent human identity on a global scale.

Alterior-wise, considering other possible global scenarios and drawing on contemporary Continental philosophy, one could imagine something like

Baudrillard's Integral Reality, a technological, *fin de siècle* system within which "we are all operators,"[9] each of us encoded into a virtuality of our own making, yet somehow still longing for a more fulfilling pact with reality. Interestingly, in Baudrillard's account of this tension between the virtual and reality something like desire still occurs, the experience at the root of transcendence and describing the epistemological condition of possibility for any critical ethics. This possibility of holding open that potential for desire, the creative gap between goal and reality, would seem closed off by this "Integral" that Baudrillard intimates. If, after Kojève we are considering the possibility of the end of history and of the turning of negating action reflexively back onto itself, as in something approximating an automatic gesture of technological integration, then we are staring into the abyss of the historical closure of desire—of human beings turned to love like young beasts, of art as natural as a spider's web or a cicada's song.[10] Thus the free act is turned back into a kind of archetypal tic, a flash of consciousness marking both the end of history and at the same time the routine automated mindfulness where reason is inflected into every artifact and institution, the beginning a stutter of a re-mythologized, hyper-rationalized history that keeps saying the name of freedom without being able to say any longer what that means.

If one considers a possible source for this idea of the integral and an integral reality in a technological, totalitarian state, then Zamiatin's *WE* is in many ways the original dystopian re-imagining of the political, patterned after book 5 of Plato's *Republic* and its mystery teaching of ideas in identity with substance and numbers conflated with reproduction, of power identified with knowledge and philosophers writing curriculum for kindergartens. In both stories, one must be aware of the tension that subsists between *eros* and *nous*, or desire and mind, where mindfulness is considered as the highest expression of desire, and yet where desire is at the same time subversive of any rationalized order of things. In *WE*, as in Plato's rationalized form of eros, there is a certain heat within the machine that destabilizes the dystopian imagination, with love out of order generating an attachment that is more particular and fleshy than any crystalline order will allow.

This is what makes Baudrillard's formulation so interestingly curious, this longing for an apparent lack, though longed for in the absence of any apparent purpose or desire for purpose. Thus, although intimating a desire for a surer pact with reality than the social contract and its modern state formations can provide, in the end Baudrillard signs off into the maelstrom of total exposure and chaotic integration with a brief manifesto: "We must defy reality as we must defy any accomplished fact." In this statement of defiance transcendence flashes from inside of the machine as in the spark igniting an automated motion, the spirit of a now increasingly machine-like planet that, as always, contains its own seeds of change within, as well as now also its seeds of self-destruction and an end of desire.

NATURE IN ITS ABSENCE

If Nature (*phusis*) "loves to hide",[11] then Nature pertains to Being.

If Nature is a presence, then Nature pertains to Being.

But would there be life worth living if there were somehow being in the absence of nature, of what is for human genetic memory the oldest of the old? What if nature suddenly quit, not at the level of the laws of gravity but in the sense of life sustained as a worldly possibility? In this line of thinking one would have to bracket a cultural philosophy of otherness, such as it is, within an associated concept of nature, of which there is much to be accounted for if the broadest, most natural context of human relations is to be understood. What this kind of philosophic study of nature and of cultural otherness could provide is insight into the limits of the technological model of subjectivity, with all of its prostheses and networks of the imagined made real, and in this sense offer a means and measure of critical self-consciousness at an historical juncture where some faculty of that kind is most certainly needed.

For it is as if, in many regards, nature is adapting to the changes wrought by historical humanity, such that the respective crises of society and of nature become difficult to discern (e.g., weather patterns are both by nature and are at the same time the products of human artifice) with the same shifts toward homogenization and integrated consequences borne out of the world we have made and the world we have inherited. So, in the face of the demand for system compliance and the homogeneity of inputs, we have extinctions of species and the streamlined downsizing of the world's store of life. Here integration in the systems of resource extraction—or what Heidegger would identify as the world put on standing-reserve in the sense of being ready to be used *qua* resources—also integrates the effects of environmental blowblack occasioned by this re-engineering of the world, forming positive feedback loops where the technological system produces its own problems, with manifest nature becoming more of a set of signs of environmental crisis than a model of balance. As the technological system starts to look more and more like an environment, nature is becoming more like technology, hardly an unexpected result of that deepening interface, such that domination is returned as it is done unto the other.

What is being manifest is a kind of corollary reaction between the human-technological and natural orders, wherein the technological ordering of things (i.e., the integrated extraction and processing of nature *qua* resources) has its reciprocal counterbalancing reactions from within the natural order, which, being a dynamic equilibrium and a superb organic system for efficient bio-integration, may overreach in stimulus response to the aberration, bringing about violent alterations of habitat, niches, surviving species, and perhaps, as reciprocal responses to environmental changes occur, even of changes to the ways human thought works. For clearly consciousness is a useful kind of tool for the worldly, adaptive animals that humans are; what is less certain and determined are what uses that thinking capacity may be put to, *qua* instrumentality. From the aggressive turn of historical human reason in early modernity toward an instrumental form of logic and an efficiency-oriented society comes this subsequent radical refashioning of the globe that human kind and all else living is launched into in late modernity. And from this arises then too these patterns of auto-rejection, intimating something deeper than the imposed system, a network that is itself still in part mediated by the third-party symbolic logics of statist and corporate forms of domination. It is as if on the barren landscapes of the technological a crystalline opening is manifest, becoming apparent as absence, where one order turns out to be a shimmering veil draped over reality, while nature makes its new claims to counter-reality, in a reaction from nature into history that is shifting the entire context for discourse from progress via mastery of human and non-human nature toward a reclaiming of the planet as our immediate present, otherwise soon to be heritage piece of historical humanity. Hence, the growing recognition of the fragile circumstances brought on via this impasse between nature and technology, out of which arises a sense of limits, indeed of the potential mortality of one's own species—quite an end-form and niche kind of thinking for a species to evolve toward. And yet, for whatever consolation it may offer, the evolution of species must always be dangerous, involving life and death, thus being the inherently vital process that it is.

CIVILIZATION AS CONSPIRACY

When people see a conspiracy, it is usually somehow seen to be in the other. It would be a rare occasion to see the conspiracy as being somehow in and of oneself—and to be in good conscience about that—that is to shape the self in conspiracy with the self. This would mean seeing an other in oneself, and one's self in another: that is the pathway to intercultural understanding.

> Could one conspire then against oneself and one's own?
> Earlier, he said he had "lost absolutely everything" on 22 July 2011, all his family and friends, therefore he understood the loss he imposed on others.
> But he said the killings were "a small barbarian act to prevent a larger barbarian act."[12]

The "larger barbarian act"—a leitmotif of projection that generally accompanies the imagining of barbaric others—would, supposedly, in this case have been an act of multicultural welcome. This was the fear behind the killings, that there would be a mixing of identities and a loss of a sense of one's own, specifically a fear of the supposed "Islamization" of Norwegian society through socialist policies of multi-culturalism. This would be the justification for acts of barbarism contra barbarism: both that the other was already there first (so, a tactic of terror against terror), and that an exclusive identity was here first, as if existing in a kind of vacuum absence of others. What may be put forward as a means of rehabilitating the paranoiac view is the realistic notion that there are always differences within, that a perfect singularity would strictly of itself be a nihil, a nothing for which the death of another is no more significant than the "cleaving of a head of cabbage or swallowing a draught of water."[13] To understand loss of life without meaning is to have participated in that same absence, with a strong feeling of purpose arising out of an experience of nihilism. Thus, otherness is inflected into identity even where differences are violently suppressed, whether by a single person or within a group of people.

At the roots of self-identity and cultural narrative is the comportment to others, which in its finer forms is known as hospitality, like in Naussica's ancient indictment spoken from an outlier island of a Hellenic civilization, Phaeacia (which would be punished just for taking in the wandering stranger, isolating it even further from the rest of humanity):

> At the world's frontier, out of all human contact.
> This poor man comes here as a wanderer,
> And we must take care of him now. All strangers,
> All beggars, are under the protection of Zeus,
> And even small gifts are welcome. So let's feed
> This stranger, give him something to drink,
> And bathe him in the river, out of the wind.[14]

That position at the margins of the Hellenic world would confirm the ethos at its center, an ethic of hospitality consisting in the recognition that gifts are welcome, that our positions in the world are precarious, and that any one of us could be the excluded.[15] For this is what made this particular civilization capable of such a resilient identity—that it could take in the stranger, make sacrifices and material gestures of welcoming, without ceasing to be what it was. Certainly the situation was helped by the geography of Scheria, land of the Phaeachians living largely "out of human contact," saved from vast multitudes of travelers, with Odysseus washing up on their shores as a kind of high water mark of Hellenic civilization. That he would be taken in by Naussica and her maidens, accepted into the court of the king, and asked to tell his story reveals a basic fraternity among these people, wanderers and settlers alike, pirates and herdsmen, homeless priests and refugees, all collectively aware of the needs of others and the essential precarity of the human condition.[16]

Such exchanges reveal quite a lot about an identity claimed; that is, the assumed identity is borne out of its relations to others. Judith Butler describes this constitutive function of comportment toward others (broadly construed on a spectrum ranging from hospitality to genocide) where an identity is formed precisely at those points of separation/connection to others:

> These are not necessarily relations of love or even of care, but constitute obligations toward others, most of whom we cannot name and do not know, and who may or may not bear traits of familiarity to an established sense of who "we" are. In the interest of speaking in common parlance, we could say that "we" have such obligations to "others" and presume that we know who "we" are in such an instance. The social implication of the view, however, is precisely that the "we" does not, and cannot, recognize itself, that it is riven from the start, interrupted by alterity, as Levinas has said, and the obligations "we" have are precisely those that disrupt any established notion of the "we."[17]

When one lets the other in, as in a gesture of hospitality, that other, even and especially in their difference, comes to belong as part of that freed identity. This would be a kind of recognition that goes deeper than an abstract formulation of I=I, or even an ennobled hypothesis of treating all humans as ends and never as means, because it can recognize difference in a person, and further see difference within an identity. For if identity is borne out of relations to others, then it will come to have a free relation with itself to the extent that it can become other than what it is.

This difference or separation from the other within an identity that Levinas identifies has much in common with Butler's emphasis on precarity as the core of the human condition and the source of our bonds to others. Both take a certain absence, expressed for example in the conditions of need, of love, of dependency, as the source of connection to others. One needs others as such, and one can love an other as such, because they are different. Levinas describes otherness as such, including comportment toward the other—say in deference, or giving heed, or listening—as the underlying thematic for our noetic, our discursive, and our ethical faculties. So knowing becomes a kind of welcoming of the other, whereby thought comes to participate in what is welcomed into consciousness. What is disavowed is a sorcerer-like summoning of the other to direct presence, magically intuiting some direct and instrumental link between the sign-forms of language and the things themselves.[18] Rather, the kind of knowing that is properly welcoming to the other is expedited by asking, listening, and paying heed to what presents itself, with the forms of knowledge thus ennobled by what is known.

This epistemological relation to the other shades into discursive questions of meaning, with the very search for meaning pre-supposing this relation to the other and the gap within which one can hold oneself out toward who others are in their differences. There is quite clearly an ethic at play in such an openness to knowledge and to meaning, with both faculties operating as modes of welcoming. It is in such a spirit that the Phaeacian King Alcinous takes in the wanderer, because "A stranger and suppliant is as dear as a brother/ To anyone with even an ounce of good sense."[19] What makes this ancient gesture of welcome more than a token of charity is that, at the rupturing of an economy of reciprocity, there is a space opened up for more than one would have bargained for, for a lack that is registered only after it has been filled in, for the other to take on a wondrous hue more substantial than the projection of self-satisfied fancies or self interests. Thus, the ethical claim of the isolated Phaeacians, distinguishing between those people who "are cruel and without right and wrong,/ And which are godfearing and kind to strangers"[20] makes way for the stranger to tell his story. The point is thus not merely to "allow" the stranger to live but to survive so that he may say who he is, telling his truth, his *muthos*, word for word and in its completeness.[21] For while there is certainly a simple economic basis for this custom of hospitality, linking together otherwise isolated island dwellers and outposts of civilization under the shared terms of precarity, beyond this material basis it is a custom that allows for the formation of identity. For these Greeks, a person is their story, and in the absence of people to tell that story to, a nameless life can hardly be called worthwhile.

If the characteristically civilizing feature of Hellenic identity is this bond of hospitality toward others—that is an identity formed within the recognition of difference—then beyond the reaches of civility there would subsist a barbarism for which others are alien nothings. In Odysseus's story these are the Cyclopes, "Lawless savages" subsisting on their own, ignoring one another and indifferent to strangers, except perhaps to eat them. This mythic story is more than an ethical test case of cultural relativism; though indeed, one could reflect upon certain truths in the Cyclopes's reasons for being suspicious of all others:

> Who are you strangers? Sailing the seas, huh?
> Where from, and what for? Pirates, probably,
> Roaming around causing people trouble.[22]

The truth in what the savage is saying is the fact that Odysseus and his men were effectively pirates, both at the siege of Troy and as they made their ways home, helping themselves to what they found along the way, like "stealing" the Cyclops's cheese for example (though one could interpret that act alternatively as a presumption of hospitality). And yet, out of their wars and wandering travels would arise a sense both of their constitutive relations to others, hosts and the hosted alike, where a certain "surplus of the other"[23] fills in the gaps between people with acts of welcoming.

What is monstrous about the Cyclops is his absolute indifference to Odysseus's entreaties to hospitality (notwithstanding the fact also that Odysseus lies about how they came to the island, saying they were shipwrecked to save the boats and the others). Odyseus's response to the monster's callous brutality is tellingly civilized, even for a pirate, when he says:

> But you're a raving
> Maniac! How do you expect any other man
> Ever to visit you after acting like this?[24]

Indeed the behavior is outrageous, however mundane the actual violence. What is striking about the violence in terms of its meaning is the sheer indifference to others. Learning the precarious situation of the supposedly shipwrecked crew, the monster calmly kills them, two at a time, reserving for Odysseus the distinction of being the last man to be killed—No-man, he calls himself, in what is a kind of empty signifier of a substitute —a *nihil* or not that is—that allows the supposed No-man to first slip past the monster discursively in order to clear the way for their physical escape. For Homer's story has the men trapped in the Cyclops's cave, where they can do little but be silent until the opportunity to escape presents itself, hiding as sheep, clinging to the undersides of the animals to sneak by the grim shepherd's blind watch.

And yet, what is so heroic about the escape is the fact that, in having escaped, Odysseus feels compelled to turn around, address the monster, and speak his own name, to say that he is not a No-man, that he has a name and a story. There is a lesson in reflexivity to this apparent need for a discursive turning back toward the monster. What it means is that the hero quite literally is his story, so that in saying what one has done, one says who and what one is, subjectively and objectively. The monster does not feel bound by any such need to account for himself before others, and so puts himself outside of the circle of others' company. He becomes as monstrous to them as he is indifferent to their lives, living in the absence of others.

An ethic of hospitality registers the inequalities inscribed into the human condition, taking heed of the fact that the world provides very different living conditions to the scattered bands of humanity. The ethic is in this sense an ethic of difference, in that it acknowledges that different people need different things, thus rehabilitating the golden rule to read: *Do unto others as those others would like done unto themselves*. For the host and the hosted are quite clearly in different positions in relation to one another. The hosted is a picture of obligation, circumscribed by unforgiving circumstances, while the host has that choice of whether to oblige or not. But if any are free, it is because they recognize this difference and are moved by the inequality of it to help change the situation of the other in need through what they do themselves.

In this sense, coming into a free relation to difference requires more than simply seeing oneself in the situation of the other, where there is always the likely danger of simply projecting one's own needs or desires onto other people. So if one lacks something that one desires for oneself, one is likely to imagine that others sharing in that absence will be similarly desirous of that missing element. When King Alcinous receives Odysseus into his home he provides for the guest with the kinds of things a king would want, with a feast and songs for the "revered stranger."[25] It takes some sensitivity on the King's part to register that the minstrel is more distressing than joy-making for the refugee. The guest is respectful in return to the host, countering that:

> Nothing we do is sweeter than this-
> A cheerful gathering of all the people
> Sitting side by side throughout the halls,
> Feasting and listening to a singer of tales,[26]

. . . but even within this shared sweetness of gathering together in song, the specific narratives of war that the minstrel sings mean something quite different for Odysseus, as the one who has himself suffered through the stories being sung. Until then in the narrative, at the cusp of Book 9 and just before entering into the dreamlike world of the hero's self-told stories, the actual name of the "revered stranger" is unknown to the hosts; that is, they know and sing the stories about the hero, though without connecting that revered name to the person himself. This dramatic situation would have Odysseus in tears, compared by Homer to those of a woman being torn from her dead husband's body and led into slavery.[27] Here there is that dramatic possibility awakened, not of some universal, singular, and homogenous identity, and of a kind of hidden, simulacra likeness of humanity, abstracted from lived conditions, but of the fact that any one of us could be in quite a different situation, that any one could potentially be transformed into an other by circumstances. It would be a curious situation indeed, having fallen entirely out of the world of men, for Odysseus to be received back into that narrated order of the commons by a song about himself. It is as if a dead man returned back in time to witness his deeds, now settled into memory and beyond his control. For there is always this element, in the case of acts of epic self-importance, of living beyond oneself, wherein the deeds and their meanings are forfeited by their agent to the witnesses and then to the story-tellers. The only way for the agent of an act to seal himself off against this dependency upon the perceptions of others is to act upon not only the episodes of the narrative but its end also.[28] The only way to entirely determine meaning exclusively according to one's own preferences and self-perceptions is to end the story as well as begin it. That is, the only entirely independent perspective is a suicidal one. For the living rather, identity subsists in relation to difference, with self-determination requiring a domain for self-presentation in the company of others.

The isolated hosts who took in Odysseus recognized the precarity of their situation, living at the margins of humanity. Homer has Alcinous twice tell the story of a prophecy that would have the Phaeacians punished for giving safe passage to a stranger, something that they would provide even still. The consequence of their hospitality would be their absolute isolation, their returning ship turned to stone and their island enclosed within a mountain. And yet for the apparent risk of becoming a hermitage from human-kind, even though they know the prophecy, the Phaeacians take in the stranger and give him what he asks for. And this is not so that he can become one of them, but first—so that he can tell his own story, saved from being a dead-man-walking shadow of a myth, disembodied from his own deeds—and second—so that he can return home. It is the distinctiveness of the hero, his story and his home, that are restored as the ends of hospitality; that is, to make the dispossessed at home in the world while listening to who they are.

It is such an ethos of hospitality to strangers, a taking in of others in the absence of any demand that they assume the identity of the host, that asserts the primacy of difference as both a condition of hospitality (i.e., if everyone enjoyed the same and equal conditions, there would be no need for such an ethic) and as its end (e.g., the King receives the revered stranger to hear his story, because a world without difference would effectively suppress identity by reducing it to a silent singularity; and it would be terrifyingly boring). These are real issues for an isolated island people. The ways in which these issues are addressed include first recognizing the other in their differences, revering the stranger, listening to their account of who they are; and second, accommodating those needs of the hosted and the differences of their condition. Thus, an ethic of difference includes both recognition and accommodation as forms of reciprocity borne out of inequality of condition.

In this sense, returning explicitly to the mass killings in Norway—the coordinated bombing of a government building and massacre of a socialist party youth camp—and reflecting on that particularly violent disavowal of hospitality to others and related policies of multiculturalism, it may be useful to locate some truth in the logic that inspired those reactionary compulsions. There is a miraculous, liberating potential to such efforts at recognition in the face of inhumanity, where it would be otherwise so very easy to dismiss the other as simply barbaric and beyond understanding. What is especially interesting in this regard is that the point of clarity is reached at the extremes of the paranoid imagination. To be clear, the common, vulgar racism behind the compulsively unethical reactions, wherein a people is reduced to an exclusive set of ethnic identifiers, signifies little beyond the abysmal ignorance of their author. However, when the ignorance reaches the fevered pitch of conspiracy, and a supposed crisis (e.g., the influences of Muslim immigration on

Norwegian society) is brought into relation with the policies and practices of the Norwegians themselves (e.g., a social democratic politics of multicultural identity and welcoming), then at those points of relation some truth shines through. For there is indeed a connection between socialist governance, youth mobilization around issues of social justice, and multiculturalism. The connection is as secure as the antipathy between violent and peaceful means is clear. What sets apart those means and links together those ends is the indebtedness to others, the sense that our being belongs in relation to others. For if we are indeed obliged to others precisely in order to assert our own identities, with who we are emerging out of how we are with others, then the dual obligations of multicultural welcoming and socialist care for the needs of others are the warp and woof that hold together a peaceful and open society. In the absence of those two dimensions of care, differences in identity and inequalities of condition have no public space within which to be recognized. What remains instead is a kind of specter of identity, already dead to the world of others, like a mad beast of a man hiding from humanity in his cave, or the abstraction of a person into a self-interested calculating automaton, a being, much like the barbarous Cyclopes, isolated by its ignorance of others.[29]

A WORLD WITHOUT BARBARIANS

> Friedrich Engels once said: "Bourgeois society stands at the crossroads, either transition to socialism or regression into barbarism."[30] What does "regression into barbarism" mean to our lofty European civilization? . . . A look around us at this moment shows what the regression of bourgeois society into barbarism means. This world war is a regression into barbarism. The triumph of imperialism leads to the annihilation of civilization.[31]

There is a sense in which barbarism is a default response for a civilization, that according to the logics of Luxemburg and Engels this famous choice between socialism and barbarism is in reality a choice between choosing or succumbing to automatic drives. Specifically it would be to choose either to take in the other according to their needs, or alternatively, to take in the other as so much fodder for the consumption mill. Indeed it may be that the instrumental appropriation of the world viewed as resources may be more productive, that imperial acquisition may be the immediately sensible response, that seeing others as simulacra of the self—as other "I's" either in the way or allied to self-interest—may be encoded into the historical condition that Engels and Luxemburg identify as bourgeois society. The dynamics of imperial appropriation do seem self-confirming in these senses, wherein a kind of obsessively blinkered view of the world forces it to be revealed as so much stuff, predigested by a perspective incapable of registering the distance and the difference between self and other.

It is at this point of closure that the barbaric pre-origins of civilization are resurrected as the expansion of a civilizational order hypertrophies into imperial excess. Here I would make a small but significant change to Luxemburg's subsequent analysis of Engels' antipodes of barbarism and socialism. She suggests that the barbarism that awaits civilization in the absence of socialism has to do with the eventual coming apart of civilization, that there is this tendency toward empire, and that from the ascent follows a decline. So, to use her illustrating example, the barbarism of the Roman empire was something that filled in the vacuum left when the empire fell. I would suggest that the barbarism was there all along in the imperial formations themselves, that the barbarity of wars of acquisition and occupation consist in the ultimate indifference toward others—viewed either as impediments or as untapped resources—a sensibility that would eventually scatter the remaining bands of humanity when the center no longer held. For the Cyclops was as barbaric when his cave was full of men as when he was alone, being isolated even in the company of others.

Even if one feels some protection in the niceties of bourgeois society, isolated from explicit violence and passively accepting the presence of others (even the Cyclops had neighbors, though with big enough backyards that they could effectively ignore one another[32]) barbarism may not be so far removed. It is not as if the veil of bourgeois indifference to others posing as politeness, the "don't step on my toes" ethic of self-isolation, has to drop in order for the barbarity within to be revealed. Rather, that specific mode of self-revealing is itself barbaric. For there is no hidden core of barbarism pre-existing social formations, but rather barbarity is revealed precisely in relations to others. For the ancient Greeks the concept illuminates a linguistic dissociation, with *barbaros* identifying peoples unintelligible to the Greeks. In that original use the word *barbaros* functions as a kind of rationalized onomatopoeia, sounding like the seemingly inarticulate sounds of the northern tribes, a "*bar bar bar*" to the civilized ear. There is no getting behind such relations, no identity pre-existing the language that gathers a people together. It is interesting that the Cyclops can speak to Odysseus, that language is useful to a being indifferent to others. For even a strictly instrumental logic has a certain logos embedded in it; and as in the New Testament story, the logos or word is in an essential sense always *with* the world and *toward* others. For that capacity to gather together the manifold of existence through expression, naming this or that, distinguishing this group of people from those others, presupposes those intervals of separation and belonging. Language is possible because there is something to speak of (i.e., language is intentional) and because there is someone to speak to, even if, as in the case of divine words of creation, the author of worlds speaks to himself.

There is some parity between the insular logic of a creator god and the radical isolation of an instrumental form of logic. In a sense, bourgeois society would be vindicated if everyone could become like their own little god, if only the bourgeois estates were large enough for the self-creation of a personal world in a vacuum empty of others. Taken to the nth degree, bourgois self-satisfaction increases to the degree that the world seems made for the self, and all others stand in as simulacra of the self—comparable in every regard—so much alike that they hardly need to be registered. One assumes that the worlds of others are essentially the same as one's own, with small aesthetic differences of taste hinting at just enough difference to cause a persistent itch of novelty, without unsettling the basic premise that the world is as one sees it, and that the world ignored may as well be invisible. This is, in the end the last great effort of bourgeoise society, sustained in the amazing capacity to ignore, with the integrity of that society dependent on the occlusion of others from view.

In the face of the now apparently democratic organization of society within the developed world, there exist latent hierarchies of domination within even politically advanced democracies, which are retarded by relations of command and obedience. All connotations of barbarism imply such relations of inequality via domination, though cumulatively tending toward being mapped out on the scale of inter-civilizational relations, with one civilization put over another. We see this in Western alliances against non-Western civilizations, which have an inherently colonizing mission, aiming to democratize non-Western societies using armed force, occupation, and political engineering, whatever the limits of political bureaucracy, occupation, and violence may be debated to be. What may be less obvious is the barbarism within, the sense of closure not only to the other outside, but toward any sense of absence within as well. Indeed, these two kinds of closures, what may be described as an absencing of absence, of life hermetically sealed off from difference, complement one another, for if one apparently lacks nothing in oneself, one has little need for others also.

ON WHAT IS MISSING

> The comparatively barbarous ages had a poetry of the ideal; they had visions of beauty, order and perfection. This age of material elaboration has no sense for these things. Its fancy is retrospective, whimsical, and flickering; its ideals, when it has any, are negative and partial; its moral strength is blind and miscellaneous vehemence. Its poetry, in a word, is the poetry of barbarism.[33]

Here in Santayana's indictment against a decline in the substantiality of our barbarism we have two poles of barbarism, one being ancient, pre-modern and "comparatively" barbarous, and the other being borne out of modernity—a barbarity that should know better. A similar, mirror division runs through the apparent identity between fascist and fundamentalist forms of violence, which would have right-wing extreme nationalist, xenophobic ideologies bearing an unintended likeness to religion-based forms of identity by exclusion. Thus, in the case of the scripted thinking behind what was an essentially anti-multi-cultural outburst of violence in the Norway massacre,

> . . . the manifesto bears an eerie resemblance to those of Osama bin Laden and other Al Qaeda leaders, though from a Christian rather than a Muslim point of view. Like Mr. Breivik's manuscript, the major Qaeda declarations have detailed accounts of the Crusades, a pronounced sense of historical grievance and calls for apocalyptic warfare to defeat the religious and cultural enemy.
> "It seems to be an attempt to mirror Al Qaeda, exactly in reverse."[34]

And yet aren't the barbarities of fascistic violence along national lines and fundamentalist religious violence both in a sense outgrowths of the critical dimension of modernity? That is, both begin with a kind of critical assessment—for example, that multi-culturalism is turning identities into outward window dressings, while occluding the very real differences between people; or similarly, that Western colonialism has flattened out the range of political alternatives, suppressing different modes of social organization by relegating them to a "backward" past. The point of rupture is also the site of attachment to the underlying ethos of modernity, namely its critical comportment toward the given, or what Hegel identifies as negation and the spirit of freedom. The point is that both fascism and fundamentalism are outgrowths of modernity in the sense of an extreme form of this negation that subsists at the core of modernity.

The fascist side of the pairing of modernity and its excesses is perhaps more obvious.[35] Zizek makes the less obvious comparison of fundamentalism and liberalism, arguing for the rather incidental medievalism of Islamic fundamentalism and its otherwise essentially modern character, namely that "Fundamentalism is a reaction—a false, mystifying reaction, of course—against a very real flaw in liberalism, and this is why it is again and again generated by liberalism."[36] By Zizek's account the weakness in liberalism consists in its parasitical relation to communal values, at once needing those values while undermining their substantial basis through the course of its development. It is on this trajectory that one can see the contradictions of liberalism manifest in corporate capitalism, with all its retrievals of authoritarian management, hierarchical organization, and oligarchical politics. For an extreme version of this manifestation, detached from the democratic political cultures that historically sustained liberalism as a set of principles, Zizek turns to capitalism under conditions of authoritarianism, seeing in that political formation glimpses of a future borne of the status quo.

> The fragility of liberalism is made clear by what is happening in China. Instead of treating contemporary China as an oriental-despotic distortion of capitalism one should see in it the repetition of the development of capitalism in Europe itself. In early modernity, most European states were far from democratic—if they were (as was the case in Holland), they were so only for the liberal elite, not for the popular classes. The conditions for capitalism were created and sustained by a brutal state dictatorship, very much like today's China: the state legalized violent expropriations of the common people which turned them into prolitarians and disciplined them in their new role. There is thus nothing exotic about China: what is happening there merely repeats our own forgotten past.[37]

Here the pattern of retrieval comes to prominence as a feature of late modernity, much as in McLuhan's media analysis and the fourth law of his tetrad, where in pushing to the extreme the new media retrieves the old, with the old order then becoming the content for the new forms of expression.[38] For in what senses can the present be anticipated to echo into the future? What will come back of what we've done, in the sense of the future returning the actions we make as consequences? It appears that what we have made collectively is a kind of barbarous system, barbaric in the sense of embodying rationality with such screaming intensity that the reasons themselves become unthinking, where even the irrational comes to happen on automatic as if by auto-rejection, with the reflexive component of modernity spinning off into a primal ending that is as technological as it is backwards, as in an ante-history at the end of history, or in more mythical terms, like Kronos eating his children to protect himself from their future.

This pattern of a retrieval of violent sensibilities is the point of rupture with and connection to what otherwise appear to be the inherently peaceful principles of modernity. These are premised upon the fundamental freedom and equality of individuals as a starting point and proceed to the construction of juridical and social orders within which individuals can enjoy that equality and those freedoms in real terms, mandated by a social contract wherein the arbiter of value is the ethical person, who is also the sole source of legitimation to state formations, there being no divine right or natural hierarchy to rule within the frame of liberal modernity.[39] Contrariwise, such reclaimed rights and hierarchies are precisely what arise as responses to the perceived flattening out of differences within modernity.

If one reflects on what the earliest critics of modernity were saying about its underlying precepts at its origins and in its early developments, then the retrieval of assumed hierarchies between peoples and of differential rights among persons would appear to answer the critique of modernity as a leveled terrain, describing conditions of life in the absence of markers for excellence or distinctions of essential difference. This critique of modernity from within goes back through Heidegger to Nietzsche and Kierkegaard and hinges on an antagonism between the exceptional person/group and the masses. For while distinctions among human beings have always existed, with awareness of that basic condition of difference being a pre-condition for the recognition of others, as well as a ground of the possibility of action, it is only in the radical hyperboles of freedom that one encounters in late modernity that the so-called "masses" can be leveled to a totally bare existence, uprooted from anything distinctive or ennobling. This is what Agamben would call "mere life," human populations consisting of the kind of world-dispossessed masses that are witnessed most clearly and shockingly in the various kinds of camps that have punctuated modern forms of social organization, where the freedoms to live as one wishes and to move around as one likes are suppressed, as in for example displaced persons camps, reserves, and, of course, in the modern state's most radical suppressions of the freedoms to life and liberty, concentration camps. If there are populations oppressed within the borders of an enclosed territory, then what they are collectively excluded from is the civilizational value attributed to those others who live within the protections of an engineered social order. If the lives of both groups, those inside and outside, receive their mandates from the civilizational order, with the insiders internalizing as self-discipline what the outsiders have imposed as the disciplines of policing, guarding, and re-habilitation (that is being made ready and palatable for settled life inside the technological bubble), then what vexes these critics of late modernity is how to find an authentic position within such a closed ideological environment. This becomes the new ethic of somehow asserting an identity that cannot be accounted for within the terms of a progressivist logic dividing the technologically and ideological settled vanguard of humanity from those living beyond the wall—those remnants of humanity whose very existence underlines the truth of Nietzsche's claim that modern man is essentially a homeless being. For in a world in which the horizon of potentials is described in terms of technology, the person is essentially unworlded, in the sense that they dwell in a world that undoes itself as world, being cosmologically detached and reduced to a map of potential perspectives and scenes, imperial world-views and ghettoized sub-cultures. That this kind of abstraction of person from world via technology would have certain re-barbarizing influences when it comes to questions of identity and of relations to others should not be surprising then, given the reductions of person to object and of populations to masses that characterize late modern political ontologies.

ON THE BARBARISM OF REFLECTION

An immanent critique arising from out of modernity can be traced back to its earliest origins, to an apprehension that a certain common sense was missing in the first reductions of reason to a tool serving individual interests. What is missing in this early modern view is any sense of what is owed to another, specifically to another who does not think in the same atomistic, calculating way characteristic of Western modernity. Even when the other may self-identify as a *homo economicus*, a well-informed consumer of worldly options, it is difficult to see how, on the basis of the isolated freedom of the individual, encountering others as if randomly bouncing into furniture in a crowded room, one could come into a meaningful relation with those others. For how does one answer what is owed to a person as such, abstracted from their communally mediated needs and rights? The concept of a social contract is an old one; for example, Plato describes democracy as an agreement among the masses of people to follow laws as protections against the wealthy or powerful few who could afford or would be able to dominate others.[40] But embedded in this kind of social contract is a natural sense of likeness, a predisposition to see society as consisting of groups of people, specifically in the instance of Plato's theory of regimes, of a society consisting of economic classes that set people apart as being rich, poor, or middle class. Here the social contract functions as means of arbitration between the divergent, yet definable interests of various groups. By contrast, the social contract in early modern political theory is a mechanism for reconciling individual interests that are taken to have no natural affiliations, no cues to belonging, no common sense overarching the "appetites and aversions"[41] that would otherwise drive a clockwork humanity to rigorous selfishness.

The social contract is thus altered, from constitutive mediation of differences, wherein a people say who they are precisely in the negotiation of their differences, to neutral social contracts serving the interests of individuals abstracted from any real association with others. At stake in the shift in meaning of the agreements that bind a community together is not only the nature of relations of person to person but the very pattern of reasoning and common sense that brings people into relations with one another and with the world. This is not simply the cost-benefit analysis informing whether to go out the door and into public but includes an appraisal of what the public use of reason is capable of. For example, is human reason able to see the purpose in natural things and thus measure the goodness of human-made artifacts in relation to those natural ends? Or similarly, is reason capable of seeing human beings in relation to some cosmological order? And even more originally, does rational discourse have a share in received patterns of thought, such as cultural expression, religion and myth; or is reason necessarily caustic to these essential fixtures of a pre-existing, ante-rational common sense?

To answer these kinds of questions of distinction would help to settle prejudices while allowing certain differences to be noticed in especial as being distinctions of integrity and hallmarks for a more rigorous and inclusive form of reason. Even to the extent that reason is shaped by a people's thinking, incubated within a history of timely rationales, and sustained by a specific language, some forms of cultural reason are more attuned, attentive and comprehensive in their sensitivity to what is. Consider, for example, a radically historicist claim that reduces all cultural values, including the value of reason, to historical valuations, made out to be essentially social constructs, whatever the cultural pretenses to transcendence or cosmogony that may be expressed by a people if they were to speak for themselves. There is an element of truth to the relativist epistemology, a part standing in for a whole, as well as a certain terror to the logic when taken to its radical conclusion. In this spirit Marx reflects that,

> Men make their own history, but they do not make it just as they please; they do not make it under circumstances chosen by themselves, but under circumstances directly encountered, given and transmitted from the past. The tradition of all the dead generations weighs like a nightmare on the brain of the living.[42]

Where is the place for reconciliation with the past within this characteristically modern frame of history as inheritance of terror? How do notions of a common good and its shared custodianship fit within this frame as distinct from modern forms of reason? For it should be clear that making one's own history in this sense means shaping the very values by which that accomplishment will be judged. And yet, this line of reasoning that has been taken up so successfully by an imperial culture of corporate modernization is incapable of turning its own logic upon itself. The radical historicist claim that all moments of thought are socially, economically and historically contingent cannot be reciprocally applied to historicism itself, which makes an ahistorical truth claim that would presumably apply to past patterns of social organization as well as to other peoples. This lack of reflexivity makes certain stark, uncompromising versions of modernity—as in corporate capitalism and revolutionary communism, rationales of progress shorn of pre-modern vestiges of their own—especially caustic to traditional societies, those with orientations to transcendence or cosmology, peoples who measure human purpose within the context of some overarching order. This shortcoming of liberal reason is reflected in the register of the person as some kind of *homo economicus*—a pleasure seeking automaton of appetites and aversions—in a basic insensitivity to cultural claims that would connect the individual within some larger ordered whole. This blind spot within modernity renders into civilizational antipathies and sources of conflict, revealing the limits of a form of reason that confuses the calculation of self-interests within purpose, while relegating the Good to an article of consumption. If there is a reason to go seeking difference as a cue to civility, the reason begins with this kind of critical assessment of where our common sense is lacking, and finds fulfillment through openness to what is beyond the self.

It is epistemologically helpful to find that juncture at which distinct elements meet, and where at the dividing line the elements share in the presence of the divided other. As ideological counterpart to early moderns such as René Descartes and Thomas Hobbes, Giambattista Vico was perhaps the original critic of modernity and its barbarous potentials, a critique that was for him directed toward the first clear expressions of technological rationality in Descartes's "Rules for the Direction of the Mind"[43] and in a Hobbesian social contract based solely on self-interest. One can take what is known as technology, that systemic assemblage of the means of worldly human freedom, as an application of Descartes's programmatic "Rules for the Direction of the Mind" onto both the world as well as onto the patterns of instrumental thought that interlink modern subjects with the objects of their making.[44] In this sense ideas and reality blend together as the mind seems to make the world. The mind (*cogito*) abstracts itself from the given world, breaks the world down, and then builds the world back up in its own image. This technological function of Descartes' resolutive-compositive method is caustic to the *sensus communis*, which consists at least partly of unquestioned collective opinions, the kinds of dispositions that incline people to bury their dead or celebrate the union of marriage. In the absence of what Vico calls these kinds of imaginative universals inherited from pre-rational, mythic modes of thought, why not compost the dead or celebrate the fragmentation of our loves?

Vico's response to modernity is to push it to its radical conclusions and to suggest how this endgame resurrects certain primitive conditions on new terms, replacing an original "barbarism of sense" with a "barbarism of reflection" that is in some ways even more spare and primordial than the original. In the radicalization of modernity one witnesses a kind of reversal wherein the hyper-rational bends back to the pre-rational, not just simply in terms of occasioning anti-modern responses but through a trimming of reason itself to a sheer instrumentality that can function in the absence of prescribed ends. What this means is that the ennobling ends that have historically been attached to the cultivation of human reason fall away as reason is reduced to a mere means by which we can get what they want, without informing us as to what is worth wanting in the first place. There are no first and last things for such a rationality recalibrated as means and as process, as in a clockwork reason that subsists in a set of mechanical rules that can be applied to thought processes irrespective of situation, circumstance, or ends.

This is the immanent critique of modernity that Vico bears out, in an indication of a certain absence at the core of the drive toward instrumentality. Where Vico diverts though from what has become a fairly widely registered critique of the flattening out effects of technological reason (as Heidegger, Horkheimer and Adorno, Marcuse, and Grant would carry on) is in his intimation that this drive toward a mode of reason turned to tool in effect produces a way of being very much like that of the myth that precedes reason. In this sense, there are for Vico two historical forms of barbarism: the first that precedes the ontological separation of subject and object, or a barbarism of sense, where humans register their senses as unmediated communion with the world; and a second barbarism of reflection that through its total objectification produces subjects who live "like wild beasts in a deep solitude of spirit and will,"[45] physically thronged together yet isolated into separate worlds.

In terms of the original barbarism, Vico adopts the approach of other state of nature theorists like Hobbes who imagine what life must have been like prior to the innovation of technological reason, that is, of human beings in their purely natural state, sans the accoutrements of civilized life. Unlike Hobbes though, Vico observes that no un-mediated knowledge of such a state would be possible, since the very being of our subjective selves—what we do and how we think—is discursively and historically encoded. The position is echoed by Rousseau when he admits that no analytical knowledge of the state of nature is really possible, to the extent that if such a state precedes the development of any forms of enculturation or social discipline, then the very forms of knowing would have to be somehow pre-linguistic and pre-rational themselves. Thus Rousseau rather lightly advances a kind of "something of the kind must be true" narrative of our prototypical origins, acting in the spirit of one "jumping over our own shadows."[46] At issue is the incredible difficulty in distinguishing what is given in our humanity and in nature and what human beings have made of both man and nature. From this arises what is a natural tendency then to transpose certain perspectives embedded in a particular social order onto others, in this case in the imagining of a state of nature pre-existing any orders of social custom.[47] So in a bourgoise liberal society, for example, property is imagined as being basic to human nature—as our primordial *in utero* where we originally connected to the world—and as a right pre-existing any social mediation. Or in a society at war the precariousness of the human condition seems to be humanity's basic element, made visible when the veil of liberal niceties drops and the limits of discourse are reached in a life "solitary, poore, nasty, brutish and shorte." And yet, all of these various accounts of what kind of human nature pre-exists the modifications of language use and of socialization are quite clearly discursive formations in themselves, with each expressing a particular modality of reason. Hobbes wants his reader to "Read thyself" in the text of the *Leviathan*, with the subject thus constituted in the terms described by the text as being basic to our humanity, namely in an appetitive and fearful psyche suited to the security state. Hobbes, like Rousseau, acknowledges that how our nature is constituted will ultimately be a function of language. And where Rousseau therefore sees a necessity "To set aside the facts" and engage in experimental, revolutionary forms of recalibrating the political subject,

Hobbes breaks through the intimate pairing of artifice and nature by a simple, sovereign fiat—*Thou shall be so*—prescribing an anxious and selfish human nature as the most stable basis for the state, with the airtight logic of an institution that can always kill those who do not fear death. Words can only go so far against this kind of self-fulfilling logic, and yet the logic itself is still a *logos*, a form of discourse embodying a distinctive grammar and vocabulary, wherein there are certain things that can said be said and certain ways in which those things can be said. What would be more honest, then, in accounting for what society has come to, where it has come from, and what it may improve toward is to first register that all of our accounting for the primordial past is itself a participation in that past, the kind of story that is shaped by what it is attempting to explain, with human nature presented as an active expression of the human spirit—that is a will to speak oneself into being that is owed to beings that preceded us.

This is the ante-rational approach to history that Vico brings to the state of nature discourses by arguing that our reasons embody pre-rational ontologies, codes via which our being comes to presence in language. The question of what constitutes human nature thus becomes an inquiry into the origins of language, an inquiry that for Vico needs to connect the substance and structure of language with a mythic point of origins at which language can be considered to be real.

This is where Vico tells a story, with his myth of origins related in various places throughout his definitive work, *The New Science*.[48] Vico's description of this ante-rational state of nature is an amalgam of historically recovered myths, focused on the cross-culturally ubiquitous feature of the flood. In his telling of the familiar story, the world comes into being from out of a watery, indeterminate unity. As the floodwaters subside one raft-full of survivors, remnants of a pre-original age, disperse and wander at will, each enjoying perfect liberty and setting off independently of the others. They live like Homer's Cyclopes, "Lawless savages who leave everything/Up to the Gods,"[49] though without gods yet to guide them, thus surviving according to the dictates of chance and spontaneous beneficence. Without any notions of distinction to set themselves apart from the natural cycles of production and consumption, these were the *indigenae*, "sons of the Earth," as Vico calls them, dwelling in such a cohesive unity with nature that each saw it as their extended body, without divisions, boundary walls, property, or laws.[50]

Out of this extensive unity arises an ontological interruption. According to Vico's cosmogonic myth, as the earth dried, it sent up "dry exhalations or matter igniting in air to produce lightening." Shocked by the sound, the sons of the Earth "raised their eyes and became aware of the sky."[51] The lightening brightened the dark sky and showed things in their outlines, distinguishing among earthly things and setting sky apart from earth. The sons of the Earth felt the thunder move through their bodies, shaking and shuddering with its noise. This was the source of their first words. For perceiving that the newly discovered sky was speaking directly to them, they echoed its sound, speaking the name of Jove and restoring their identity with the now divided world through speech.

This was for Vico the origin of language from out of the barbarism of sense, wherein a kind of direct affinity was formed between words and their referents, whether in terms of an identity of likeness, where by onomatopoeia words are hitched to things as their signs, or by an identity of effect, wherein language exercises a magical power over the world to which it relates. This is the domain of myth and sympathetic magic, where by outward likeness and the look of a thing some element of an essence is indicated, given in the self-presentation of a being. And by this mythic self-revealing quality of beings, certain handles on reality are created, liminal junctures at which the two elements—referent and symbol—share in one another, each bearing the imprint of the coterminous other. So reality takes on a shape and self-presentation—like the mountain range traced out against darkness by lightning—as the language or *muthos* inherits a corporeality from what first moved it into being.

This pattern of inheritance from being into language is for Vico then translated through the development of language, with the forms of being given to a poetic imagination that would become the material for reflection. Thus the original acts of imagination via which the world of nature is appropriated as sign, stamped onto the sensorium in a moment of wonder and realized for its significance (i.e., for the sons of the Earth, the storm must mean something), become the source of a vocabulary joining form and substance.

In this seeking for the roots of reason in the poetic imagination, Vico carries out a rather unique etymology that would trace back the Greek *logos* to origins in *muthos*, with myth encoded in every reason, by noting a simple and basic meaning of both words as story, or even more simply, as word. This suggestive parity runs counter to the thrust of the history of ideas from Plato on through Christianity and into modernity, with a break marked between *logos* as true speech, being loyal to the law of non-contradiction, and *muthos* as mere speech, unbound by internal consistency precisely by its spontaneous affinity to truth understood as mimetic effect.

This is the ontological basis of what Vico would call the barbarism of sense, with perception so enthralled by the world as presence that it seems to light up magically with meaning. This identity between expression and the expressed would be the basis of the original cultural formations in Vico's new mythology, his *New Science*—common traits supposedly shared by all civilized peoples—namely reverence for the dead, observing the bonds of marriage, and giving worship. That is to say that barbarism is encoded into a civilization precisely in its universalizing features, its most consistent reasons, and its unquestionable forms of common sense. There is a sense in which these most basic reasons for a people's being, what Vico calls its "imaginative universals," have a certain mythic non-sense included at their core, an otherness at the heart of reason itself, with the original code then transmitted in recombinant instantiations via a people giving an account of themselves over time in their culture, language, history and politics.

This pattern of mythic association attaches a sense of substance to the historical world in which man makes himself. For across the gulf of cultural differences, certain common traits find their expression. Even if peoples across the world do not say precisely the same things, they say them in certain shared modes of expression, as in, for example, saying something about respect for a person in the specific cultural modes of funeral rites; or saying something about the worshipful character of humanity in its propensity to religious sentiment; or reflecting on devotion to an other encountered in the most immediate sense, that is as a beloved. These basic cultural continuities for Vico point to common origins in myth, more specifically to a mythic sensibility within which the world is seen as an extension of human being, in a sense moving through the person when they give voice to it. Thus at first heaven was no higher than the mountain tops,[52] and the first peoples imagined the world itself invested with spirit, being animated and sacred in its very being. There is real intelligence in the seemingly ignorant state of mind that will not distinguish self from world—as in a kind of preemption of disbelief—in that it describes a basis for common understanding that is not the result of a particular civilizing code having been passed from one peoples on to another but is grounded rather in the spontaneous receptivity of people to the world conceived as extended, sacred body.

How this mythic sensibility is undone, broken into separate domains of free subjectivism and material objectivity, is in many ways the history of modern reason. By reason one understands what beings are for; that is, *What is their reason for being?* both in terms of subjective freedom and objective materialities. Or there is reason as rationale, as in how one gives an account of why a being is in terms of what has become of it. Or one addresses the less philosophical, yet still constitutive reasons that pertain to the purpose of a society and its means of organization. Or there is reason at the level of the logic of a specific discourse, like the code that makes a text readable and structures the thoughts coming out of it. At all of these junctures where reason touches down in order to structure a reality, there is flexion between that structured reality and the code that shifts things into order, acting like a lever onto the sheer givenness of sensory experience so that trains of events are informed and given purpose, objects put into objective relations with one another, and subjects identified by their perspective within the altered reality (all so-called realities are altered and must be in order to accommodate subjective freedom in a world of objectifications). At these points of flexion between code and reality, where reality is torqued into order, changes in the patterns of thought and in reason itself become possible, opened up where the mode of thinking reveals a slightly off realization, a kind of surplus to the intents of reason. Here reason comes to bear most significantly in its unintended consequences, wherein the consequent surplus of reality reveals the limits of the logic employed, grounding a history of ideas in the acknowledgment that each moment in that history is a kind of revelation of the purpose of reason in terms of what it cannot do. Reason is in this sense then ultimately a mode of self-consciousness, wherein the self is known by its limits.

If we are rational subjects motivated primarily by the intimation of our own limits, in the most extreme variant by our consciousness of mortality, then why then should one care for the being of another? Why should each individual subject not obsess instead over his or her own security? This is the social contractarian position that Vico confronts in Hobbes, in modernity contra the first ante-modern position, the latter of which has the aim of restoring a *sensus communis*—an affinity deeper than interest in one's own self-preservation will allow.

What occurs at that precise historical juncture, between pre-modern traditions of thought, which were invested in myth, and modernity, which is a divestiture of myth, is the re-shaping of reason—from *cosmos* to instrumentality. For within myth the *communis* has its basis in the *cosmos*, and it is this sense of a part within a whole that a strictly procedural rationalism cannot grasp, being limited to language forms that bifurcate symbolic orders into subject and object. What remains of the whole becomes a kind of technological function, a partial remembering of a sense of cosmos within a system for which, symbolically and functional-wise, the whole is a cumulative function. The specter of integration that haunts the new shape of reason recasts the feeling of union as abstract, artificial, and empty, such that what brings people together becomes source of their alienation also.

To take a reciprocal approach to the dilemma, reclaiming the spirit of union precisely in the experience of difference, just what is it then that is the source of difference, and further, how can differences be known? If knowing a being as what it is implies both synthetic and analytic capacities, that is the capacities to both gather together a specific sense of a being as what it is, as well as to distinguish a being *qua* what it is from out of a surrounding environ, then what identifies a being participates in both its analytically distinct features, and in its affinity for what is around, in a kind of likeness that operates across categories and the differences that define them.

The substance of this likeness within difference subsists in the form of an identity whose identity can be identified with its other, that is an identity with the other as constitutive element. In this sense, within the civilizational imperative toward integration, the recourse to barbarism is both a natural reaction to the excesses of technological modernity, as well as being a radical realization of the potentials of a modernity unbridled by constitutive differences, their myths and sources of belonging. For however closed off and inherently "different" an identity may be, identifying itself in terms of being "other than" some supposed others, that irreconcilable difference is itself sourced in specific relations to others. That is, the difference itself is a relational rather than self-subsistent concept. Everything that is intrinsic to the relational subject is thus also extrinsic to it, outside of itself, much like an avatar's identity is transparent to the outside, while acting as a mask to what is within. Contrariwise, the confused disposition toward difference as something that can somehow be sealed off, while one's own identity is assumed to be transparent to oneself and one's own, limits the fulsome expressions of identities—cultural, religious, political, and linguistic—by freezing them into caricatures of a people; while in reality what remains of a people is for them to remain at odds with themselves, held together by mutual suspicion. Rather, in an open society constituted in terms of a free relation to difference, patterns of recognition would be extended to multicultural identities where the other within is not merely incidental to a dominant culture, acceptable to the extent that niceties would allow within current corporate-liberal society, but where those constitutive relations make a people who they are collectively, each of them with their differences intact.

ON EXCLUSION

> What is the relation between politics and life, if life presents itself as what is included by means of an exclusion?[53]

The existence of differences remains the basis of possibility for both the political and for life as such—what Agamben calls "mere life"—with the drive to differentiation being active in the motives, means, and ends of actions as it is in life's simple yearning after its own sweetness of being.[54] For it is always *this* specific life that desires its own continuation, or that is attached to the particular life of another as being dear in and of themselves and for their own sake. These are the sorts of things that cannot be exchanged, unattached to a note of currency that can somehow stand in for the thing of value, it being valuable precisely inasmuch as it is different, and, therefore, valuable outside of identity with some neutral third term. Thus, at the origins of political involvement one registers a sense of the importance of the actors' own dear opinions, while at its ends, in action and in discourse it is the differing perspective of the other that is recognized and included, brought into a dialectical relation with the self so as to belong within an identity. This is what gives the political and the possibility of free association their qualities of inherent fragility, in that our accomplishments in the collective sphere always depend upon others, especially inasmuch as it is a care for difference that is the specific thing of concern.

The specific value of difference as such is to provide the conditions of possibility for political life, for the gathering of a plurality would not be necessary if its starting point were one of unity and spontaneous concord. In terms of life itself, the possibility that life can become other than what it is, that is, that the forms of life can change through growth, mutation, and differentiation, provides for the ontological sustainability of life in general. In this sense, the fact that there are different morphologies present within individuals, between species, and over time through evolution belies the notion that there somehow exists a baseline measure of "bare life," some kind of undifferentiated proto-biological substance from which all living beings have been gathered. The very existence of life, in any of its multitude of forms, necessarily stands out. One feels this baseline quality of the distinctiveness of life in the sweetness that can attend to it, as in the basic desire for its continuation, whatever its specific circumstances may be. When one registers a feeling of pleasure at simply being alive, it is at a particular moment, in one place, when a person's environ converges into a particularly enjoyable set of conditions, coupled with the propensity to enjoy them. That kind of charmed disposition necessarily requires attunement to specifics as mode of engagement. The sense of bliss in the experience of simply living may indeed be unqualified, as in crossing over the boundaries of whatever norms are mapped onto the teeming vivacities of life in order to gain some social purchase on reality via abstraction, but because they are without qualification does not mean that they are indistinct or otherwise neutral to identities. The sheer sweetness of being is unqualified because it stands out, such that through joy some specific sense of goodness comes to prominence.

It is not, therefore, that the political quality of human life—consisting in the paired capacities for action and free association—is somehow an addendum to an underlying condition of merely being alive as a human being, like a set of clothing that the naked body puts on to become a public person, but rather that being political is an expression of life, to the extent that action and free association with others are what life is directed toward. All of nature inclines toward this kind of grouping together and focusing of individuated efforts, which stand out because there is something to stand out from, as in an individual arising out of community, in identity formations that are borne out of an other within.

The full meaning of this essential relation of difference to identity resonates through those moments of life when its precariousness is a clear and present reality. That one would purposefully risk or expend one's life for a cause amplifies what is an essentially enframed dimension of the human condition, determined as it is between endpoints that set the limits and possibilities of a person's life and actions. For while one can to some extent choose how one's life will end (if one were so inclined to choose death, as in an act of bravery or of suicide), one cannot choose one's parents.[55] And in those rare cases where such a decision to mark that mortal end deliberately arises not out of relegation to the conditions of one's existence but rather out of determination that the current conditions shall not stand, with a life put in the way of things carrying on as they always have, as if fate would deliver a self to its circumstances as a river carries a leaf along, there a life gets to decide just what it will be in the fullest sense possible. This is the heroism of self-sacrificing action, set apart by the courage to come to an end willingly and with a purpose freed from the tragedies of reciprocity and retribution.

The free act has a certain quality of being, as Arendt describes, "infinitely improbable," occurring outside of the deadlocked cycles of blame and transferred identities where the alien other is seen in terms of a kind of gnostic separation of self from self—as in barbarians with whom there is no speech, relation, or interdependence possible. Alternatively, the possibility of the "infinitely improbable" being enacted is predicated upon seeing the other as having a share in one's own identity. Not only does this initial perspective of recognition free collective action from the compulsiveness associated with a divided identity—the kind of inchoate self-assertion that tilts at its own windmills—it also gives the actors themselves purchase on the power of the other, in the sense of identifying with one's own weakness, cueing on that fulcrum of imbalanced interdependence as a potential source of movement.

Thus Arendt points to the peaceful revolt of subjects against imperial or tyrannical governments as way of short-circuiting the imbalanced power of state over person, effectively defusing the state's claim to sovereignty as an excuse for violence by demonstrating the capacity to organize in the absence of coercion:

> Popular revolt against materially strong rulers, . . . may engender an almost irresistible power even if it forgoes the use of violence in the face of materially vastly superior forces. To call this "passive resistance" is certainly an ironic idea; it is one of the most active and efficient ways of action ever devised, because it cannot be countered by fighting, . . .[56]

There is this persistent sense in which, for Arendt, a genuine action stands out from the field of otherwise determined relations, being free of the imperatives of retribution for and escalation of antecedent causes. Reciprocally, the free act engenders no set response. Indeed, under the most pressing of circumstances, as in, for example, a campaign of mass disobedience against civil government, the response may be an impasse of options, with the state left with no adequate course of action, except to stop acting, reduced to the bare sovereign power not to respond. It is this end of the state in inaction, announced quietly by the retreat behind palace walls that opens up the miraculous dimension of coordinated action, in a kind of collective heroism that mirrors the teleological quality of individual heroics. So Arendt has it that,

> . . . human essence—not human nature in general (which does not exist) nor the sum total of qualities and shortcomings in the individual, but the essence of who somebody is—can come into being only when life departs, leaving behind nothing but a story.[57]

Seen under contemporary circumstances, it is curiously paradoxical that this once-defining feature of what is civilized—the training of the passion for life toward self-sacrifice—has now become the very hallmark of those who are considered barbarian within neo-liberal discourses. What supposedly makes the new barbarians so irrational and ruthless in their use of violence is this disavowal of life. This is what short-circuits the opposing liberal logic, putting the suicidal barbarian so far outside of its range of hearing. And from the other perspective, what is apparently barbaric is precisely what neo-liberal modernity offers up as manifest benefits, namely material prosperity and a Hobbesian/Lockean social contract standing in as common bond. It is easy to see how liberal ideology could appear vulgar to a passionately devout person, and reciprocally how the inclination to self-sacrifice could be vulgarized into cheap rhetorical effects, with senselessness being proof of purity, such that civilization undoes itself through what is a radicalization of its core principles.

For taken to the extreme of liberalism's anthropocentric bias, there would be no place for higher goods in public discourses, short of the flattened-out code of efficiency. From an anthropocentric, that is, a strictly human perspective and materialist epistemology—tethered to what eyes and ears could come to know—the view is always the view from a particular, individual person, whereas notions of the collective or a common good must have some perspective of the other included, reaching out in words that are more than instruments of the senses or birddogs for the desires of each to their own. For how does one feel connected with the world, if not by how the other comes into contact, by how welcoming that contact is, by how much a part of oneself one feels the other to be?

This kind of ethos of hospitality that is expressed in acts of welcoming offers a point of connection between what are perhaps the most archetypal of differences among people, traceable to the distinction between a community and its strangers. How these two groups either get along or are at enmity with one other is a matter corresponding to the quality of this constitutive relationship. Thus, it becomes a question of whether others are freely accepted within a community, free to travel and work, or whether others are occluded in their specificity, taken to mean just difference as such, as in a pure other, *nihil*, and mask for hidden inhumanity. When it comes to a matter of whether one kind of people can live with another kind, whether made different by stature, skin, language, or culture, there is that breaking point at which the other appears to fall away, when there is no talking with them, no trust with them, no shared understanding. Something of the humanity of the other changes when this occurs: they become inscrutable *How could they have done this?* and shockingly alien to what the mass of humanity understands itself to be.

This is where the question of some supposed motives of barbarity gets waylaid, whether by orientalist attitudes or in seeking to understand the other objectively in terms of the interests that may enframe the barbaric act, historical sources of hatred and injustice, or psychologies of sociopathy. What is missing in these terrorized questions is that they don't get beyond the potential of the act, as if interests, history, or psychology could provide some likely reason for barbaric behavior. There is a streak of sublimated racism in the shocked horror of the West recoiling, *Why did **they** do this to **us**?* There is this desperate yearning in the second-by-second news feed to get into the conscience of the other, to register how conscience itself could make that kind of pitched demand of a person, that is, how the conscience could effectively nullify conscience. What reason will not find, terror can produce, where what remains is an absence of trust in others, supplemented by a thin gnostic faith mistaking paranoia for enlightenment. For considering matters of security—that benchmark starting point for relations with others—the question is ultimately an ontological one, with the other being either a part of an identity or considered somehow as external to it. Is the other welcomed or are they feared? There is always something of a missing element, a projected image relocated and mapped out onto the other, where what is lacking in one's own culture is filled in by that difference. Whether this absence is registered reflexively, seeing the other in the self, or unconsciously, seeing identity in terms of a difference *from* . . . , still those differences have their constitutive place within every identity. It is when the others are little understood that they become the perfect *tabula rasa* for imagined identities, seeing ourselves in terms of our self-made others.

In attempting to heal the community by excising the barbarian, keeping up the appearances of civilization in the face of trespasses on it, the communal identity is already infiltrated. It is in those anticipatory gestures by which a civilization undoes itself, amplifying the effects of whatever barbaric act it was, with the supposed outside finding its way into the envelope of insular identity, in what then becomes an institutionalization of the original violence. How this closure to the other comes to be a matter of principle, of institution, and of culture has the encounter echoing into a people's future, indicating the barbaric potentials of what will come out of lost opportunities for understanding.

NOTES

1. Martin Heidegger, *Pathmarks*, "On the Essence and Concept of *Phusis* in Aristotle's *Physics* B.1"

2. Alain Badiou, "Nature—Poem or Matheme?", *Being and Event*, Oliver Feltman translation (London: Continuum, 2005), 121-29.

3. G. W. F. Hegel, *The Phenomenology of Mind*, 284.

4. G. W. F. Hegel, *Reason in History*, 20.

5. Thomas Hobbes, *Leviathan*, chp. 13.

6. See Janet Gross Stein, *The Cult of Efficiency*, (Toronto: Anansi, 2002). It is not by accident that Stein ends her critique of efficiency in systems' organization with a chapter on terrorism, written in the ramping up of the War on Terror, or that she counterpoints social democratic governance and the managerial ethos. Nor is it an oversight to diminish any existential importance to those radical acts of violence. For it is better to have a dull drive to efficiency than an existential one, better to witness a global crime than a global war.

7. George Grant, "A Platitude" in *TECH-NO-LOGY AND EMPIRE: Perspectives on North America,* (Toronto: Anansi, 1969), 142.

8. Steven Pinker, *The Better Angels of Our Nature: A History of Violence and Humanity* (New York: Penguin, 2012).

9. Jean Baudrillard, *The Intelligence of Evil or The Lucidity Pact,* Chris Turner translation, (Oxford: Berg, 2005), 44.

10. Alexandre Kojève, *Introduction to the Reading of Hegel*, James H. Nichols translation (Ithica: Cornell University Press, 1969), 159fn.

11. Heraclitus, fragment 17.

12. "Breivik trial: Phone delay 'caused more Breivik deaths" *BBC news*, 23 April 2012 http://www.bbc.co.uk/news/world-europe-17809617.

13. G. W. F. Hegel, *The Phenomenology of Mind*, 605.

14. Homer, *Odyssey*, Stanley Lombardo (trans.) (Indianapolis: Hackett, 2000), 6.210-16.

15. There is an inkling of textual/historical evidence to support this thematic comparison, with some scholars suggesting that the Phaeacians whom Homer describes may have in fact been Norsemen, with Odysseus' travels taking him outside of the Mediterranean to the North Sea and back. See Armin Wolf, "Mapping Homer's Odyssey" Institute for Neohellenic Research, *Eastern Mediterranean Cartographies* (2004), pp. 309-34. This potential convergence of identities, historical and mythic, offers an interesting rejoinder to the prejudice that would frame identities formed within such far-flung hinterlands as Scandinavia as being somehow isolated or set apart from the rest of the world. No people is isolated unto itself except in the ignorance of others, and even that lack of awareness contains within it intimations of interdependence.

16. Judith Butler, *Frames of War*, 43.

17. Judith Butler, *Frames of War*, 14.

18. Emmanuel Levinas, *Totality and Infinity*, 92.

19. *Odyssey*, 8.591-2.

20. *Od.* 8.621-2.

21. *Od.* 3.94, 4.324.

22. *Od.* 9.246-8.

23. Levinas, *Totality and Infinity*, 97.

24. *Od.* 9.247-9.

25. *Od.* 8.589.

26. *Od.* 9.5-8.

27. *Od.* 8.566-74.

28. Hannah Arendt, *The Human Condition*, 193-4.

29. *Od.* 9.112.

30. The reference is to Engels' *Anti-Dühring*, (Moscow: Progress Publishers, 1947), II.ii, 203.

Why bourgeois society specifically? Because a thinking society always stands at a precipice, precariously balanced against the default position implied by ignorance, with its intentions working out beyond the original intents.

31. Rosa Luxemburg, *The Junius Pamphlet*, Chp. 1 http://www.marxists.org/archive/luxemburg/1915/junius/ch01.htm.

32. *Od.* 9.112.

33. George Santayana, *Interpretations of Poetry and Religion*, (New York: Harper, 1957), 168-9.

34. Steven Erlanger and Scott Shane, "Oslo Suspect Wrote of Fear of Islam and Plan for War," *New York Times*, July 23, 2011. http://www.nytimes.com/2011/07/24/world/europe/24oslo.html?ref=europe.

35. See Gene Veith's *Modern Fascism*, (St. Louis: Concordia Publishing House, 1993).

36. Slavoj Zizek, *Living in the End Times*, (London: Verso, 2010), 154.

37. Ibid.

38. Marshall McLuhan, *The Laws of Media*, (1988).

39. Clearly there are other frames available, especially when it comes to end-considerations of political legitimacy, with other models becoming increasingly apparent, as in for example still-existing monarchical states, newly arising corporate-communist states, oligarchies, empires, tribal formations, etc.. From the liberal perspective of a development-oriented approach that reflexively considers its own conditions of being in relation to other possible frames, it is thus safe to say that:

> Legitimacy is a complex issue. . . . An important part of the research agenda on conflict and development is to gain a better understanding of the dynamics of legitimacy—including, vitally, non-western forms of legitimacy such as revolutionary legitimacy, or embedded/historical legitimacy—and the manner in which it plays into causes of conflict and of recovery.

Bruce Jones, Molly Elgin-Cossart and Jane Esberg, "Pathways Out of Fragility: *The Case for a Research Agenda on Inclusive Political Settlements in Fragile States*" (Discussion paper, 2012).

40. Plato, *Republic*, 359a-c.

41. Hobbes, *Leviathan*, 118.

42. Karl Marx, "The Eighteenth Brumaire of Louis Bonaparte," 97 in *Karl Marx and Frederick Engels: Selected Works*, (Progress Publishers; Moscow, 1968).

43. In *Descartes: Philosophical Writings*, translated and edited by Elizabeth Anscombe, and introduced by Alexandre Koyré (London: Nelson, 1969). Koyré, from whom Alexandre Kojève inherited his famous Hegel seminar and some of his ideas, has an introductory essay in this volume in which he praises Descartes as a crucial turning point in the supposed progress of philosophy. Koyré emphasizes the importance of the themes of freedom and mastery in Descartes, saying that both forms of relation begin with acts of negating the given.

44. Donald Phillip Verene marks this origin of technological process thinking in Descartes, making a critique of technology via the critique of early modernity. See "Persons in a Technological Universe" in *Globalization, Technology and Philosophy*, Tabachnick and Koivukoski (ed.) (SUNY, 2003).

45. Giambattista Vico, *The New Science*, Bergin and Fisch translation (Ithica: Cornell University Press, 1948) §1105.

46. Hannah Arendt, *The Human Condition*, (Chicago: University of Chicago Press, 1958), 10.

47. "And all of them, constantly speaking of need, avidity, oppression, desires, and pride, have transferred to the state of nature ideas which they acquired in society; they have spoken of savage man, but depicted social man." Jean-Jacques Rousseau, "Discourse on Inequality Among Men", in *The Essential Rousseau*, 144.

48. Giambattista Vico, *The New Science*, §62; §369-73; and §1097.

49. *Od.* 9.105-6.

50. *The New Science*, §370.

51. *The New Science*, §377.
52. *The New Science*, §64.
53. Giorgio Agamben, *Homo Sacer: Sovereign Power and Bare Life*, Daniel Heller-Roazen (trans.) (Stanford: Stanford University Press, 1995)
54. Aristotle, *Politics*, 1278b25-30.
55. It is curious that the corresponding terminus of natality is not such a consideration for existential thinkers after Heidegger. The concept does regain some prominence in his student's work, described by Arendt as an essential aspect of political action in the sense of making a beginning.

Yet, in what has become the traditional, disproportionate ideological emphasis on death as the basis of the political (as in hard sovereigntist claims of a supposed right of the state to execute its own populations) in the main and in technological society the very real frailty of the human condition is crowded out by human-made uncertainties, with nature remapped by technology and with a sense of precariousness within the natural balance occluded by the paranoias of a world-system with built-in kill-switches, as in a system that would negate itself, or a world that would self-destruct.

56. Hannah Arendt, *The Human Condition*, 201.
57. Hannah Arendt, *The Human Condition*, 193.

Bibliography

2002 National Security Strategy of the United States of America.

Agamben, Giorgio. *Homo Sacer: Sovereign Power and Bare Life*, Daniel Heller-Roazen translation. (Stanford: Stanford University Press, 1995).

Agamben, Giorgio. *Means Without End: Notes on Politics*, Binetti and Casarino translation. (Minneapolis: University of Minnesota Press, 2000).

Anderson, Benedict. *Imagined Communities: Reflections on the Origin and Spread of Nationalism*, (London: Verso, 2006).

Arendt, Hannah. *Between Past and Future: Six Exercises in Political Thought*, (New York: Viking Press, 1963).

Arendt, Hannah. *On Revolution*, (London: Penguin, 1990).

Arendt, Hannah. *Origins of Totalitarianism,* (New York: Harcourt, Brace & World, 1950).

Arendt, Hannah. *The Human Condition*, (Chicago: University of Chicago Press, 1958).

Aristotle, *Politics,* Ernest Barker translation. (Oxford: Oxford University Press, 1958).

Badiou, Alain. *Being and Event*, Oliver Feltman translation (London: Continuum, 2005).

Barthelemi, Jean-Jacques. *The Travels of Anacharsis the Younger in Greece During the Middle of the Fourth Century before the Christian Aera*, 2nd edition translated by Duc de Nirvernois. (London: Vernor and Hood and Lackington, Allen, and Co., 1798).

Bataille, Georges. *DOCUMENTS*, 7, 1929. Dominic Faccini translation in *Undercover Surrealism: Georges Bataille and DOCUMENTS*, Dawn Ades and Simon Baker (ed.) (Cambridge: MIT Press, 2006).

Baudrillard, Jean. *The Intelligence of Evil or The Lucidity Pact,* Chris Turner translation. (Oxford: Berg, 2005).

Berger, Jonathan and Thomas Terry, *Canoe Atlas of the Little North*, (Boston Mill Press: Boston, 2007).

Butler, Judith. *Frames of War: When Is Life Grievable?*, (London: Verso, 2010).

Campbell, Joseph. *The Masks of God: Creative Mythology*, (New York: Penguin, 1968).

Cassirer, Ernst. *Language and Myth*, Susan K. Langer translation. (New York: Dover, 1946).

Cassirer, Ernst. *The Philosophy of Symbolic Forms, Vol. 2: Mythical Thought*, (New Haven: Yale University Press, 1996).

Cavafy, C.P. *C.P Cavafy: Collected Poems*, Edmund Keeley and Philip Sherrard translation. Edited by George Savidis. (Princeton University Press, 1992).

Descartes, René. *Descartes: Philosophical Writings*, translated and edited by Elizabeth Anscombe, and introduced by Alexandre Koyré. (London: Nelson, 1969).

Eliade, Mircea, *Cosmos and History: The Myth of the Eternal Return*, William Trask translation. (New York: Harper, 1954).

Eliade, Mircea. *Myth and Reality*, William Trask translation. (New York: Harper, 1963).

Engels, Friedrich. *Anti-Dühring*, (Moscow: Progress Publishers, 1947).

Gadamer, Hans Georg. *Reason in the Age of Science*, Frederick Lawrence translation, (Cambridge: MIT Press, 2001).

Grant, George. "A Platitude" in *TECH-NO-LOGY AND EMPIRE: Perspectives on North America,* (Toronto: Anansi, 1969).

Habermas, Jürgen. *The Divided West*, (Cambridge: Polity, 2006).

Hannah Arendt, "*Zur Person*," interview. http://www.youtube.com/watch?v=dsoImQfVsO4.

Hegel, G. W. F. *Reason in History,* Robert Hartman translation. (New Jersey, Prentice Hall, 1997).

Hegel, G. W. F., *Phenomenology of Spirit*, Baillie translation. (New York: Harper, 1967).

Heidegger, Martin. *Pathmarks*, William McNeill translation. (Cambridge: Cambridge University Press, 1998).

Heraclitus, *Fragments: The Collected Wisdom of Heraclitus*, Brooks Haxton translation. (New York Viking, 2001).

Hobbes, Thomas. *Leviathan,* (London: Penguin, 1985).

Homer, *Iliad*, Stanley Lombardo translation. (Indianapolis: Hackett, 1997).

Homer, *Odyssey,* Richard Lattimore translation. (New York, Harper and Row, 1999).

Huang, Yong. "The Ethics of Difference in the *Zhuangzi*", Journal of the American Academy of Religion, *(2010) 78* (1): *65-99.*

Irigaray, Luce. *Speculum of the Other Woman*, Gillian C. Gill translation. (Ithica: Cornell University Press, 1985).

Jonas, Hans. *The Gnostic Religion*, (Boston: Beacon Press, 1991).

Jones, Bruce, Molly Elgin-Cossart and Jane Esberg, "Pathways Out of Fragility: *The Case for a Research Agenda on Inclusive Political Settlements in Fragile States"* (Discussion paper, 2012).

Kalevala, Aili Kolehmainen Johnson translation. (Hancock: Book Concern, 2015).

Kant, Immanuel. *Critique of Pure Reason.* Norman Kemp Smith translation. (Blue Press, 2007).

Kant, Immanuel.*Perpetual Peace*. Lewis White Beck translation. (Pearson, 1957).

Koivukoski, Toivo. *After the Last Man: Excurses to the Limits of the Technological System.* (Lanham: Lexington Books, 2008).

Kojève, Alexandre. *Introduction to the Reading of Hegel*, James H. Nichols translation (Ithica: Cornell University Press, 1969).

Kovel, Joel. *The Enemy of Nature: The End of Capitalism or the End of the World*, (Zed: London, 2007).

Lévinas, Emmanuel. *On Escape,* (Stanford: Stanford University Press, 2003).

Lévinas, Emmanuel. *Totality and Infinity,* Alphonso Lingis translation. (Pittsburgh: Duquesne University Press, 1969).

Luxemburg, Rosa. *The Junius Pamphlet*, http://www.marxists.org/archive/luxemburg/1915/junius/ch01.htm.

Machiavelli, Niccolo. *The Prince*, Robert M. Adams translation. (New York: Norton, 1992).

Marx, Karl, "The Eighteenth Brumaire of Louis Bonaparte," in *Karl Marx and Frederick Engels: Selected Works*, (Progress Publishers: Moscow, 1968).

Marx, Karl. *Economic and Philosophical Manuscripts of 1844*, in *Writings of the Young Marx on Philosophy and Society*, Loyd Easton translation. (Garden City: Double Day, 1967).

McLuhan, Marshall, Eric. *The Laws of Media*, (Toronto: University of Toronto Press, 1988).

Mouffe Chantal. *On the Political*, (London: Routledge, 2005).

Friedrich Nietzsche, *Beyond Good and Evil*, Walter Kaufman translation. (New York: Vintage, 1989).

Nietzsche, Friedrich. *The Gay Science*, Kaufmann translation. (New York: Vintage, 1974).

Nietzsche, Friedrich. *Untimely Meditations*, R.S. Hollingdale translation. (Cambridge: Cambridge University Press, 1997).

Partridge, Eric. *Origins: A Short Etymological Dictionary of Modern English*, (London: Routledge and Kegan Paul, 1958).

Pinker, Steven, *The Better Angels of Our Nature: A History of Violence and Humanity,* (New York: Penguin, 2012).

Plato, *Republic*, Allan Bloom translation. (USA: Basic Books, 1991).
Plato, *Symposium,* Seth Benardete translation. (Chicago: University of Chicago Press, 2001).
Plato, *Theaetetus*, Fowler translation. (Cambridge: Harvard University Press, 2006).
Rousseau, Jean-Jacques. "Discourse on the Origin and Foundation of Inequality Among Men" in *The Social Contract and Discourse on the Origin of Inequality*, anonymous translation. (New York: Washington Square Press, 1967).
Said, Edward. *Culture and Imperialism*, (New York: Vintage, 1993).
Santayana, George. *Interpretations of Poetry and Religion*, (New York: Harper, 1957).
Shakespeare, William. *A Midsummer Night's Dream*, (New Haven: Yale University Press, 1965).
Smith, Adam. *The Wealth of Nations*, (New York: Modern Library, 1937).
Sontag, Susan. *Regarding the Pain of Others*, (New York: Picador, 2003).
Sophocles, *Oedipus Tyrannus,* Francis Storr translation. http://www.perseus.tufts.edu/hopper/text?doc=Perseus%3Atext%3A1999.01.0192%3Acard%3D1329.
Stein, Janet Gross. *The Cult of Efficiency*, (Toronto: Anansi, 2002).
Strauss, Leo, "Political Philosophy and History" in *What Is Political Philosophy?* (Chicago: University of Chicago Press, 1959).
Strauss, Leo. *Natural Right and History*, (Chicago: University of Chicago Press, 1953).
Thucydides, *History of the Peloponnesian War*, Richard Crawley translation. (Vermont: Everyman, 1993).
Veith, Gene. *Modern Fascism*, (St. Louis: Concordia Publishing House, 1993).
Verene, Donald Phillip. "Persons in a Technological Universe" in *Globalization, Technology and Philosophy*, Tabachnick and Koivukoski ed. (SUNY: 2003).
Vico, Giambattista. *The New Science*, Bergin and Fisch translation. (Ithica: Cornell University Press, 1948).
Virilio, Paul. *Open Sky*, Julie Rose translation. (New York: Verso, 1997).
Virilio, Paul. *Speed and Politics: An Essay on Dromology.* Mark Polizzotti translation. (New York: Semiotext(e), 1986).
Voegelin, Eric. *Science, Politics and Gnosticism*, (Washington: Regnery, 1997).
Voegelin, Eric. *The New Science of Politics*, (Chicago: University of Chicago Press, 1987).
Wolf, Armin. "Mapping Homer's Odyssey" Institute for Neohellenic Research, *Eastern Mediterranean Cartographies*, pp. 309-34 (2004).
Wollstencraft, Mary Shelly. *Frankenstein*, (Boston: Bedford, 2000).
Yeats, William Butler. "The Second Coming" http://www.potw.org/archive/potw351.html.
Zizek, Slavoj. *On Belief*, (New York: Routledge, 2001).
Zizek, Slavoj. *Living in the End Times*, (London: Verso, 2010).

Index

About the Author

Toivo Koivukoski is associate professor of political science at Nipissing University and Director of the Nipissing University Peace Research Initiative. He is the author of *After the Last Man: Excurses to the Limits of the Technological System* (Lexington Books, 2008), and co-editor with David Tabachnick of a series of books on classical regimes theory, including *On Oligarchy* (2011), *Enduring Empire* (2009), and *Confronting Tyranny: Ancient Lessons for Global Politics* (Roman and Littlefield, 2005), along with *Globalization, Technology and Philosophy* (2004).

www.ingramcontent.com/pod-product-compliance
Ingram Content Group UK Ltd.
Pitfield, Milton Keynes, MK11 3LW, UK
UKHW012042150926
475560UK00001B/141

9 780739 189993